5th Biennial Conference of the Interdisciplinary Society for Quantitative Research in Music and Medicine (ISQRMM 2019)

Published in the Journal of the Interdisciplinary Society for Quantitative Research in Music and Medicine

Rockville Centre, New York, USA
7 - 9 June 2019

ISBN: 978-1-7138-1357-6

Printed from e-media with permission by:

Curran Associates, Inc.
57 Morehouse Lane
Red Hook, NY 12571

CURRAN ASSOCIATES INC. proceedings.com

Some format issues inherent in the e-media version may also appear in this print version.

ISBN: 978-1-7138-1357-6

Printed by Curran Associates, Inc. (2020)

For permission requests, please contact ISQRMM at the address below.

ISQRMM
C/O David O. Akombo, Ph.D.
Dean, Faculty of Culture, Creative and Performing Arts
P.O. Box 64
Bridgetown BB11000
Barbados, West Indies

http://isqrmm.org

The *Journal of the Interdisciplinary Society for Research in Music and Medicine* is a peer-reviewed journal presenting scholarly research into music's connection and application in medicine. Articles are submitted from scholars who have presented at the society's biannual conference, as well as from the field. Inquiries regarding submission should be directed to contact@isqrmm.com

Additional copies of this publication are available from:

Curran Associates, Inc.
57 Morehouse Lane
Red Hook, NY 12571 USA
Phone: 845-758-0400
Fax: 845-758-2634
Email: curran@proceedings.com
Web: www.proceedings.com

Interdisciplinary Society for Quantitative Research in Music and Medicine

Proceedings Journal 2019

President's Forward

The study of the beneficial effects of music in medicine is a discipline that has spurred an interest among scholars since the dawn of the 18th century. Scholars from many spheres of human development have been intrigued but especially those from the humanities and biomedical sciences. Researchers in the 19th and 20th centuries continued to embrace these studies with scholarly and academic vigor paving the way for those in the 21st century to explore it even deeper. The mission of the Journal of the Interdisciplinary Society for Quantitative Research in Music and Medicine is to propel this scholarship with vigor into the 21st century and provide a global readership of peer reviewed articles that explore the unanswered questions regarding the efficacies of all forms of music on the human body, mind and soul. The Journal of the Interdisciplinary Society for Quantitative Research in Music and Medicine welcomes qualitative research as well as quantitative research methodologies and encourages interdisciplinary formats. The Journal encourages the authors to utilize modes of research that are clearly defined, explained and relevant. The Journal also welcomes Master's or Doctoral theses in any related fields.

The topics of historical significance impinging upon qualitative research on music and healing, especially as revealed in its manifold and remarkable manifestations—throughout Ancient Greece in the 18th century inspiring scholarship around ancient civilizations—has never been candidly and fairly examined in the lenses of the contemporary scientific models. To this end, there a dearth of research spanning the global cultures that use music as non-phrenological medicine are worthy of this examination. Presently, the only sound sources of information on the culturally diverse aspects of music as medicine are anecdotes, which provide eyewitness accounts of individual music healers. In addition to the quantitative studies in the field, the anecdotal repository of evidence, as recorded in treatises of healing rituals, add credence to the historical implication of music as a healing agent.

The Journal of the Interdisciplinary Society for Quantitative Research in Music and Medicine contains cutting-edge research that balances its content with readership that is interdisciplinary in scope, while using eclectic models to present the research findings. These make the journal both scholarly and informative for all readers regardless of specialization. We are proud of ISQRMM community, enthusiasts, and erudite scholars that make up our global research clientele.

Musically Yours,

Dr Akombo

David O. Akombo, Ph.D.

Co-Founder and Board Member

Editor's Forward

Music is a profound and transcendant medium. There is no person who is not readily influenced by music; elevated, descended, enthralled, or enraged, it reaches our soul. It's profound effects on the human mind and body have historically been seen as medicinal or toxic.

Herein, our journal seeks to bring together the proceedings and work of a heterogenous and disparate bevy of scholars and practitionars who have all come to the conclusion that music can be considered an aspect of medicine. Their approaches are all different, their insights are varied, but they all share this conclusion that music can effect the mind and the body as a type of medicine. They bring together theory and practice, with empirical observations, clinical trrials, and experiments.

Their results reinforce and reproduce their observations and conclusions. This is science – the reproducibility of result and cross corroboration derived of multiple scholars and multiple methodologies. The goal of the Interdisciplinary Society for Quantitative Research in Music and Medicine, is to provide a platform for this type of research to be encouraged, shared, and expounded upon.

I am honored to edit the Procedings Journal and first hand have met and learned from the body of contributers. As editor, it is my sincere hope that the reader can gain insight and inspiration from the collection of papers, and hopefully join us at our next conference and participate in the shared journey of academic exploration in the dimensions of music as medicine.

Steve Jackowicz, Ph.D.

Editor and Board Member

Contents

An Unsculpted Block of Time:

A Humanistic Approach to Using Music Therapy with Acute, Adult Psychiatric Groups

Jon Reichert

Abstract

In the US, a distinguishing feature of facilities where acute, adult psychiatric patients receive treatment is the relatively rapid patient turnover. Due to frequent admissions and discharges, membership in a unit's groups can vary daily. Given the fluctuating membership, as well as the evolving mental states of group members, every group session is a unique experience. The question arises: How does a music therapist prepare for a group session where the composition and mental states of its members are unknown? This article provides a conceptual framework, based on humanistic principles that can be used in helping to determine what music therapy methods to use to effectively provide therapeutic benefit to a group's members.

In the US, there are several aspects of the *acute* psychiatric care environment that distinguish this patient population from the overall mental health population. First, when acute patients are admitted to a facility, it is because they have experienced some type of psychotic episode and/or psychological crisis, which often causes them to present with unique behaviors and challenges. A second distinguishing feature is the facilities where acute psychiatric patients receive treatment, which are "locked", meaning that patients are not able to leave a facility until their treatment team determines that they are ready for discharge. The fact that patients cannot leave a facility under their own volition can contribute to agitation and feelings of frustration and anxiety, which can manifest themselves in a patient presenting with threatening behavior towards other patients and staff.

A third distinguishing feature of the care provided to acute psychiatric patients is the relatively brief treatment period. In the US, the average length of stay has been estimated to be in the range of 7-13 days (Lee, Rothbard, & Noll, 2012). A fourth aspect, as seen in Park's (2011) research, which is related to the brief treatment period, is the relatively rapid patient turnover. Due to frequent admissions and discharges, a unit's group membership can vary daily. Additionally, the mental state of individual group members can vary each day, depending upon how they are responding to their prescribed medications.

Given the fluctuating membership of a unit's music therapy group and the evolving mental states of the group's members, every group session that is conducted is truly a unique experience. Clinicians who work with this population are confronted daily with the question: How does a music therapist prepare for a group session where the composition and mental states of its members are unknown? Nachmanovitch (1990) stated, "As stone is to a sculptor, so time is to a musician. Whenever he gets up to play, the musician stands there facing his own unsculpted block of time" (p. 31). When working with acute, adult psychiatric patients, this author views the start of each group session as though the group is being presented with an unsculpted block of time, and I am always curious and fascinated to see what musical experiences the group creates out of that block of time. In that vein of thinking, this article provides a conceptual framework that music therapists can use in determining what methods to use to effectively provide the greatest therapeutic benefit to a group's members.

Wheeler (1983) designated three psychotherapeutic levels of music therapy practice: activity, reeducative, and reconstructive. As Wheeler pointed out, a key difference between the activity level and the other two levels is that, "the *therapy* in music activities comes largely from the involvement of the person in a therapeutically oriented activity rather than from talking about this involvement" (p. 8). According to Wheeler, the attainment of insight is required only at the reeducative and reconstructive levels. At the reeducative level, the therapeutic focus is on helping the patient function at a higher level, and at the reconstructive level the objective is to alter the basic personality structure of the patient. Although the three level framework is applicable to different client/patient populations, Wheeler's research focused on how it applied to adult psychiatric clients.

Wheeler's (1983) research was based partly on the work of Wolberg (1977), who had developed a framework for types of psychotherapies that included supportive, reeducative, and reconstructive levels of therapy. Wheeler stated, "Supportive therapy has as its goal to restore individuals to an emotional equilibrium so that they can function as closely as possible to their normal levels" (p. 9). The focus at this level is on reducing symptoms, ego strengthening, and developing better "mechanisms of control" (Wolberg, p. 68). The supportive level of therapy is particularly applicable to acute psychiatric patients because it can be used both with those who have a reasonably sound ego structure and require therapy for only a short time period to return to normal functioning, as well as with patients who are so severely incapacitated that the probable best outcome of treatment is symptom alleviation. Supportive therapy is also applicable for acute psychiatric settings because therapists in these facilities, at least in the US, typically are not afforded the necessary time to work with patients at a more intensive level.

To date, the supportive/activity level of music therapy has been portrayed in the literature primarily as a means of achieving behavior modification in patients. Wheeler (1983) explicitly discussed the use of activity therapy as a means of achieving behavioral change, portraying the therapist as an authoritarian figure who directs the sessions. Wilson (2005, p. 111) discussed activity therapy as being "particularly well suited for reducing inappropriate behaviors…", while Scovel and Gardstrom (2005, p. 117) stated that, "activity-based therapy is aimed at helping the client reach observable, measurable goals…"

Alternatively, this article provides an approach for working at the supportive/activity level of therapy with groups of acute, adult psychiatric patients that is based on a humanistic theoretical orientation, incorporating principles of Yalom, Creative Music Therapy (CMT), Rogers, and positive psychology. The author would also like to propose that within the supportive/activity level, there are three levels of possible patient engagement which roughly align with three of the main methods of music therapy delineated by Bruscia (2014): improvisatory, re-creative, and receptive.

Although composing/songwriting is another method of music therapy designated by Bruscia, this author has not included it as a level of patient engagement within the supportive/activity level of therapy because songwriting requires a level of mental functioning that many acute psychiatric patients do not possess. Regarding the use of music versus verbal processing in the therapeutic process, Bruscia (1998) has described the concept of music *as* therapy as occurring when the therapeutic process happens either entirely through music, or when verbalization occurs, but only to interpret or enhance the music experience. Alternatively, Bruscia describes the concept of music *in* therapy as occurring when the therapeutic process happens either equally through music and verbalization, or when the process occurs primarily through verbalization. While it is conceivable that all four music therapy methods could be used in either music *as* therapy or music *in* therapy approaches, composing/songwriting is the one method that would most likely be utilized in a music *in* therapy approach. Although some acute psychiatric patients do have the capacity to participate in songwriting experientials, this method is likely more effectively used at the reeducative and reconstructive levels of therapy.

An underlying premise of the three levels of patient engagement is that, in a group setting, those musical experiences that provide opportunity for the highest level of creative involvement also possess the highest potential for cathartic, therapeutic change. Of the three levels of patient engagement, improvisatory experiences possess the relatively greatest potential for creative involvement while receptive interventions typically possess the relatively least. A particular patient's ability and willingness to participate will determine the experientials that he or she will engage in, and as mentioned earlier, this can vary on a daily basis depending in part on the patient's mental condition and how he or she is responding to their medication(s).

The goals primarily targeted using this humanistic approach include increasing patients' self-esteem, interpersonal skills, and socialization skills. Yalom (1983) discussed the importance of setting realistic goals for acute psychiatric patients. He wrote that for inpatient groups, goals of effecting major changes in character structure are overly ambitious and can be anti-therapeutic. "To set goals of amelioration of acute psychotic behavior for the small therapy group is unrealistic and will, almost certainly, result in the therapy group being, and being considered, ineffective" (Yalom, p. 53). Yalom and Leszcz (2005) described achievable goals for inpatient groups to include engaging patients in the therapeutic process, decreasing isolation, affirming self-esteem, and decreasing hospital-related anxiety. They noted, "... it is imperative that the therapist create in the group an atmosphere that members experience as supportive, positive, and constructive" (Yalom and Leszcz, p. 488).

As mentioned earlier, improvisatory experientials possess the greatest potential for creative involvement, particularly non-referential improvisations. Improvisational music experiences allow each group member to develop their own, unique common ground for creative expression and communication, and allow each patient the flexibility to come in and out of that common ground as they choose. There is no right or wrong level of participation. Patients are accepted for who they are. The over-arching aim of the experientials is to provide patients with a positive experience and a sense of accomplishment.

The use of improvisation with psychiatric patients has been discussed by Stephens (1983), who pointed out that through the use of percussion and melodic percussion instruments, virtually all patients are able to participate, regardless of musical training. Citing CMT as an example, she noted, "Through spontaneous musical interaction, the relationship between therapist and client is developed. The improvised music provides a medium for communication, psychological organization, self-realization, and creative expression" (p. 30). She also stated, "An important element is that of an equal relationship in musical terms. It is not the therapist, a trained musician, playing beautiful music for the client, but both struggling with rhythm and non-rhythm, loud sounds and soft sounds, fast sounds and slow sounds" (p. 41).

Solli (2008) detailed a case study of his work using improvisation with a man with schizophrenia in a psychiatric inpatient unit. Solli pointed out the importance of recognizing and utilizing patients' own musical preferences. The potential for conflict between the approaches of music *as* therapy and music *in* therapy was highlighted when, during a session, Solli unsuccessfully attempted to engage the patient in verbal dialogue, and the patient finally responded back to him, "Shut up and play!"

Aigen (1996) discussed the concept of musical intelligence, the "constellation of perceptual, aesthetic and expressive abilities" (p. 13) that can allow a patient to escape the confines of his or her pathology. Aigen noted that while some patients can fully access and express their musical intelligence through musical structures, other patients require an improvisational approach in order for their musical intelligence to become fully manifested.

During a session, the group leader is certainly responsible for maintaining a safe environment and procedural order. However, during the group improvisations, the role of the group leader is primarily non-directive in nature. The leader's primary role is to play in a manner that invites and supports group members' playing and vocalizing, creating and maintaining group cohesiveness, and creating a positive environment. This non-directive role is based on Rogers' person-centered approach, which embodies a deep sense of trust in the group's ability to develop its own potential (Corey, 2012). The group members are challenged to struggle to express themselves, and out of this struggle they have a basis for learning how to trust themselves. According to Rogers, when the leader effectively conveys genuineness, positive non-judgmental acceptance, and empathy towards group members, therapeutic change occurs. This non-directive, non-authoritarian role also helps mitigate potential tension with group members who are defiant of authority.

The use of group improvisation incorporates ideas associated with positive psychology, specifically Csikszentmihaly's (2008) concepts of flow and optimal experience. Optimal experiences (the best moments in our lives) are not the passive, relaxing times. Rather, they occur when a person's body or mind is being stretched in a voluntary effort to accomplish something. Optimal experiences are something we make happen. They occur when people are in a psychological state of *flow* – a state of mind in which someone is so involved in an activity that nothing else seems to matter. According to Csikszentmihaly (2008, p. 41), "Flow helps to integrate the self because in that state of deep concentration, consciousness is unusually well-ordered." Similar to the idea of musical intelligence allowing a patient to escape the confines of his or her pathology, one aspect of flow is that, while it lasts, a person is able to forget all the unpleasant aspects of life.

As mentioned earlier, a patient's ability and willingness to participate will determine what experientials they engage in. Not all patients are willing and/or able to participate in group improvisations, but they will participate in re-creative interventions such as sing-alongs. Although typically not providing the same potential for creative expression as improvisations, re-creative interventions can provide an effective vehicle for patients to express themselves and achieve the goals of enhancing self-esteem, interpersonal skills, and socialization skills.

For those patients who are not willing and/or able to engage in either improvisatory or re-creative interventions, receptive methods, such as music listening groups of patient-preferred music, can also help patients achieve the aforementioned goals. Having a group of peers applaud a particular song choice or discovering that other peers share a particular musical interest can help a patient feel a bit more comfortable in a living environment that can sometimes be unsettling and frightening. Receptive interventions can also sometimes transform into re-creative or improvisatory experientials when patients choose to sing along and/or dance to the music being listened to.

In conclusion, although the supportive/activity level of therapy has traditionally been discussed in terms of a behavioral theoretical orientation, it can also be thought of in terms of a humanistic theoretical orientation. Within the supportive/activity level, there are three levels of possible patient engagement: improvisatory, re-creative, and receptive. This discussion has hopefully provided the reader with a conceptual framework that can be used in designing experientials for use with groups of acute, adult psychiatric patients that will both attract patient participation and be therapeutically effective.

References

Aigen, K. (1996). *Being in music: Foundations of Nordoff-Robbins music therapy* (Monograph). St. Louis, MO: MMB Music Inc.

Bruscia, K.E. (1998). *Defining music therapy* (2nd ed.). Gilsum, NH: Barcelona Publishers.

Bruscia, K.E. (2014). *Defining music therapy* (3rd ed.). University Park, IL: Barcelona Publishers.

Csikszentmihalyi, M. (2008). *Flow: The psychology of optimal experience.* New York, NY: HarperCollins Publishers.

Corey, G. (2012). *Theory & practice of group counseling* (8th ed.). Belmont, CA: Brooks/Cole, Cengage Learning.

Lee, S., Rothbard, A.B., & Noll, E.L. (2012, Sept.). Length of inpatient stay of persons with serious mental illness: Effects of hospital and regional characteristics. *Psychiatric Services, 63*(9), 889-895. doi:10.1176/appi.ps.201100412

Nachmanovitch, S. (1990). *Free play: Improvisation in life and art.* New York, NY: Jeremy P. Tarcher/Putnam.

Park, A.A. (2011). *Theoretical orientations applied to music therapists working in adult psychiatric inpatient settings* (Master's thesis). Retrieved from http://digitalcommons.molloy.edu/etd

Scovel, M.A., & Gardstrom, S.C. (2005). Music therapy within the context of psychotherapeutic models. In R.F. Unkefer, & M.H. Thaut (Eds.), *Music therapy in the treatment of adults with mental disorders: Theoretical bases and clinical interventions* (pp. 117-132). Dallas, TX: Barcelona Publishers.

Solli, H.P. (2008). Shut up and play! Improvisational use of popular music for a man with schizophrenia. *Nordic Journal of Music Therapy, 17*(1), 67-77.

Stephens, G. (1983). The use of improvisation for developing relatedness in the adult client. *Music Therapy, 3*(1), 29-42.

Wheeler, B.L. (1983). A psychotherapeutic classification of music therapy practices: A continuum of procedures. *Music Therapy Perspectives, 1*(2), 8-12.

Wilson, B.L. (2005). Transitioning from institution to community. In R.F. Unkefer, & M.H. Thaut (Eds.), *Music therapy in the treatment of adults with mental disorders: Theoretical bases and clinical interventions* (pp. 104-116). Dallas, TX: Barcelona Publishers.

Wolberg, L.R. (1977). *The technique of psychotherapy* (3rd ed., Pt. 1). New York, NY: Grune & Stratton.

Yalom, I.D. (1983). *Inpatient group psychotherapy.* New York, NY: Basic Books.

Yalom, I.D., & Leszcz, M. (2005). *The theory and practice of group psychotherapy* (5th ed.). New York, NY: Basic Books.

Music and Art Therapy to Build Posttraumatic Growth with Women in the Psychiatric Mental Health Setting: A Pilot Study

Laryssa M. Creswell, Deirdre Cogan, Kate Stanley

Abstract

Trauma treatment for women with severe and persistent mental illness often is not addressed due to the fear of psychological decompensation (Sanders, n.d.). Through music, art and psycho-education these women can learn to differentiate sensations, imagery, and responses connected to past traumatic events from their current realities. The purpose of this study was to determine whether the Empowerment Guide improves post- traumatic growth for women with severe and persistent mental illness, substance abuse and trauma histories; also defined as co-occurring disorders. The study was a single-subject quasi-experimental design. A combination of music and art therapy interventions were used for this study. Improvisation, lyric analysis and receptive techniques such as music relaxation and listening were used in conjunction with cognitive re-structuring imagery and projective drawings. Four out of ten eligible participants consented to participate in the study. The participants met the following criterion: (1) severe and persistent mental illness, (2) trauma history, (3) minimum age of 18 years old and (4) currently housed on a long-term unit to allow opportunity for more psychological stability. Attrition factors lead to only 2 participants completing the pre-& posttest. The limitations of the design approach and the unique qualities of this population restricted our ability to determine whether the empowerment guide improved post-traumatic growth in women with co-occurring disorders. The results of this study inform music and art therapy clinicians of the need for a mixed method design to better inform gender specific treatment.

Introduction

The rapid acceleration of women entering the criminal justice system has been alarming, and the prevalence of female recidivism gives cause for concern. Throughout the late eighties and mid-nineties, arrest rates for women steadily grew by 200% (Pomeroy, Kiam, & Abel, 1998). Women are now the fastest growing demographic within the system. Covington (2007) states that nearly one in a hundred women in the United States is a detainee in a correctional facility. Three quarters of female inmates serve prison time following two or three prior arrests (Alemagno, 2001). However, it is also important to note crimes committed by women differ from those committed by their male counterparts. By contrast to male offenders, female offenders are arrested most often for non-violent crimes due to domestic disputes and economic hardship. Many women arrestees are caught in a perpetual cycle of drug use, prostitution, theft or fraud (Covington, 2007).

The pervasiveness of trauma and its aftermath are noted to be high in cost and a harmful public health concern. The traumatic incidents bring about profound burdens for the survivor, their families, the community at large and the public health agencies charged with providing treatment services (SAMSHA, 2014). Vogt (2018) reported women to be more susceptible to sexual assault and sexual abuse as a child over and above men in a recent survey. The author described results from community research studies of post-traumatic stress disorder (PTSD) as an adverse response to traumatic incidences for 9.7% of women compared to 3.6% of men in the United States. This occurrence is 2 to 3 times more for women than men (Vogt, 2018). Although trauma research has garnered increased attention in the literature over the years, it continues to be disregarded for those with severe mental illness (SMI) receiving treatment (Mauritz, Goossens, Draijer & van Achterberg, 2013). The National Institute on Drug Abuse (NIDA) (2018) reported guidelines have been developed to encourage federal agencies to include women and analyze gender and sex differences in research. The needs of women with severe mental illness, trauma and substance abuse disorders often go unnoticed (Creswell, 2016). Specifically, there is a necessity to research and determine effective components of treatment for women with severe mental illness (Creswell, 2014).

Mauritz et al. (2013) argue the lack of attention to this population is due to the commonalities for trauma related and SMI symptoms, and the trepidations clinicians feel to not exacerbate symptoms for persons with SMI. Van den Berg et.al, (2016) found trauma focused treatment for individuals diagnosed as severely mentally ill with psychosis did not worsen, revictimize nor have adverse responses. Instead, found there was significantly less adverse responses compared to those on a waitlist. Efforts to conduct research on the complexity of women's lived experience with substance use, trauma and severe mental illness have steered organizations to include gender-specific treatment to address their unique needs (Gunn, Sacks, & Jemal, 2016).

Dass-Brailsford & Safilian (2017) stated approximately 80% of women pursuing treatment for substance use disorder, also report physical and sexual assault histories. Those traumatic memories are often stored in the unconscious and in the body; therefore, when activated these memories can move quickly into automatic perceptions, feeling and behaviors (Sareen, 2014). When this cycle repeats itself, women may associate their symptoms and their maladaptive behavior with negative core beliefs about themselves. Using the creative arts as an effective tool has gained significant consideration to prevail over obstacles, encourage self-awareness, self-expression and learning (Gussak & Ploumis-Devick, 2004). Re-processing traumatic memories through structured creative arts therapy groups encourages women to reconsider, even challenge, their existing assumptions about themselves and about the world around them. Because of the unique societal roles and psychosocial needs of women, research has shown conventional treatment methods to be less effective and more diverse treatment options such as acupuncture, hypnotherapy, music therapy and art therapy to be more beneficial (Aletraris, Paino, Bond-Edmond, Roman & Bride, 2014).

Review of Literature

Music Therapy, Addiction & Severe Persistent Mental Illness

Research has shown that 75% of those seeking treatment for substance abuse disorders are commonly diagnosed with psychiatric disorders as well. Unsuccessful treatment of addiction often occurs due to the sparse availability of treatment services, the vast range of motives for use and the complexity of engagement in treatment (Megranahan & Lynskey, 2018). While pharmacological and psychosocial treatments have advanced significantly, many with SMI continue to struggle to progress with traditional treatments (Gold et. al, 2013). The authors noted music therapy as a beneficial form of practice to encourage expression and socialization for low motivated clients with SMI. Moreover, stressing the importance for active involvement of research to continue development as an evidence-based therapy (Gold et. al, 2013). Koelsch (2015) found that music serves a purpose for participants and involves social cognition, synchronization of actions, increase social cohesion and can lead to empathy.

Grocke et. al (2014) stated music therapy has been recognized as an effective form of treatment for individuals with SMI for over 10 years. The use of music provides an outlet for self-expression, and increases socialization, motivation, optimism and encouragement. Gold et. al, (2005) reports results of music therapy from a Cochrane review as improving socialization, mental and global state. The objective of music therapy is described as seeking to assist people with SMI to foster relationships and deal with issues that are otherwise difficult to express with words. Rolvsjord (2015) found adults in mental health care settings engaged in music therapy were active in seeking the value of the therapeutic experience and not just merely passively enjoying the musical abilities of the therapist and captivated by the musical interventions. Additionally, participants chose specific music to fit themes, regulate emotion, commitment to the therapeutic relationship, and brought musical objects from their everyday life showing their active engagement in the process.

Music therapy from a resource-oriented perspective uses the strengths rather than problems of the person with SMI and builds on a relative knowledge base of the therapeutic process, positive psychology and empowerment (Gold et. al, 2005). Williams, Dingle, Jetten & Rowan (2019) explored the effectiveness of individuals with SMI participating in arts-based groups. Specifically, to determine whether group singing versus creative writing formed a sense of belonging and purpose as well as cultivate an empowered identity. The authors found both the creative writing and group singing improved overall well-being. McCaffrey, Edwards & Fannon (2011) found the focus on strength building and the supportive nature of music therapy fits well with the recovery approach in mental health settings. Music therapy offered chances for individuals with SMI to socialize, increased engagement and enhanced retention in the therapeutic process.

Music Therapy & Trauma

The use of verbal processing of traumatic events often becomes troubling, invasive and overwhelming. Music therapy has proved to be advantageous for trauma survivors in order to create a felt sense of safety, rediscover their healthy identity and engage in treatment for a greater period (Carr, et al., 2012). The fundamentals of music therapy create a locus that can address the unique needs of women, such as physical safety, the feeling of inclusion, security to express emotions and validation seeking, which are all essential when a woman is in recovery (Gardstrom, Klemm & Murphy, 2017).

Techniques such as musical improvisations rattle unconscious matter to be converted into the conscious; making music therapy a safe way to expose, handle and heal traumatic experiences (Amir, 2004). Moreover; music can go directly to the memories and emotions, as well as accessing the parts of the unconscious that are unsafe and threatening for the person (Amir, 2004). Due to this notion, music therapists possessing the tool of music have a significant advantage and grave responsibility, when treating those who have been traumatized (Swallow, 2002). It is the time-ordered structure of music that creates a means for a person to experience themselves in time and moment to moment

(Sutton, 2002). The author goes on to say music establishes the opportunity to express the chaos and fragmentation experienced by a person with trauma histories and make what is formless into form and bring about the mourning and loss of their former self (Sutton, 2002).

Several neurotransmitters such as dopamine and serotonin are released to increase musical gratification (Hodges, 2016). Music may play a role in offering therapeutic value to the wounded amygdala, which likely stimulates the release of hormones such as endorphins and adrenocoticotrophin (ACTH) (Sutton, 2002). Consequently, as this occurs it supports the feeling of being grounded within the body at the same time; making processing or re-experiencing more accessible when engaged musically (Sutton, 2002). Hodges (2016) reported musical emotions are widely embodied throughout the brain. An initial inquiry into neuroimaging and musical emotions found increased blood flow to the amygdala and other sections of the brain occurred during what is known to be the chill response.

Carr et. al (2012) reported outcomes of their randomized control study which proposed that music offered a way for self-regulation and arousal stimulation, in addition to successfully creating a safe environment to take up the issue of avoidance. By gaining control of arousal and avoidance, the space is created for re-experiencing to be addressed. Hodges (2016) stated the use of familiar music to the client's culture will stimulate various regions of the brain versus non culturally familiar music. Moreover, this familiar music activates the prefrontal cortex when the memories are brought forth and are autobiographical in nature. Ahonen (2016) discussed healing factors for participants in the process of Group Analytic Music Therapy for trauma. The initial stage being voicing and witnessing the trauma, grieving the trauma and its aftermath. Once this occurs, Ahonen (2016) states the restructuring of the world as it was known, restoration of trust, regaining a sense of reality and lastly reintegrating the personality can begin to take place.

Art Therapy and Trauma Treatment

A review of literature reveals an abundance of articles discussing the efficacy of art therapy in the treatment of trauma exposure and Post Traumatic Stress Disorder (PTSD). However, there are remarkedly few resources on this topic for individuals living with Severe Mental Illness. Many concur that traumatic memories are stored as visual and sensory experiences or implicit memories. While explicit memories are easily retrieved, implicit memories by pass language and thought. They are encoded in the limbic system and in parts of brain which are beyond conscious awareness (Talwar, 2007). Difficulties emerge when implicit memories are not linked to explicit memories. Rather, when individuals struggle to make meaning of felt sense and their emotions that are disconnected from which they arose. (Rothchild, 2000). Art therapy can be used as a tool to access the parts of the brain where implicit memories are stored (Meckums, 1999). Art making is an integrative, sensory mode of expression which involves all five senses; therefore, it facilitates the retrieval of sensory memories (Steele 1997; Steele and Raider 2001).

Malchiodi (2001) notes that art expression connects implicit and explicit memories. When exploring traumatic experiences through drawing, one must think and feel simultaneously, creating a bridge between affect and cognition. The author also notes that during her interviews with children from violent homes, art making helped to diffuse anxiety, enabling children to provide more detailed and coherent stories (Malchiodi, 1997; 2001). Creating artwork can also provide the emotional distance necessary for one to make narrative meaning of traumatic memories (Smiejsters, 2008). Implementing basic tenets of mindfulness combined with the art making process can activate the body's relaxation response (Malchiodi,2001). Working while in a relaxed state can expand the perimeters within which an individual can tolerate exploration of painful memories and emotions (Seigal, 1999).

Approaches employed with those struggling with acute PTSD may require highly structured methodologies. Some authors warn that exposure to traumatic material via highly charged, graphic imagery can lead to or reliving of traumatic experiences (Barnes, Cohen., & Rankin, 1995). While providing short-term relief, this process can undermine the longer-lasting benefit of self- regulation and symptom management. Some suggests that a focus on skill building through cognitive re-structuring combined with expressive techniques helps to minimize the risk of decompensation (Barnes, Cohen., & Rankin, 1995).

Creativity combined with cognitive re-structuring is seen as a means through which a traumatized individual can reconsider or challenge assumptions, building a more sustainable recovery (Barnes, Cohen., & Rankin, 1995). Johnson (1999) contends that there are numerous parallels between paradigms underlying creative arts therapies and cognitive-behavioral therapies. The author notes growing trends away from exposure therapy towards containment techniques in the treatment of trauma survivors. The author speculates this shift arose from ongoing concerns about re-traumatization (Johnson, 1999). If art therapists are attuned to widely held concerns about the re-traumatization, it is incumbent upon art therapists to develop approaches that help to mitigate this risk.

Art Therapy and Addictions Treatment for Individuals Living with Severe Mental Illness

The America Art Therapy Association (2014) acknowledges art therapy as an effective intervention in managing addictions. Numerous articles underscore therapeutic benefits of using art therapy with individuals who are undergoing addictions recovery. The medical model plays a central role in the amelioration of symptoms in substance abuse recovery due to physiological stress associated with detoxification, and underlying changes that substance abuse has on circuits in the brain (Stein & Friedman, 2001). Focus is often placed on art therapy as a complimentary form of treatment that involves social and emotional factors. The sequela of substance abuse is complex and is disruptive to the many facets of one's life; therefore, a need for more holistic approaches focusing on all aspects of a person is indicated (Donovan, Molina & Kelly, 2009).

Art therapy can facilitate exploration of the interconnections between thoughts, feelings and behaviors of individuals struggling with substance abuse issues (Holt

& Kaiser, 2009). Moore (1983) suggests combining the elements of cognitive-behavioral principles helps to promote self-awareness, decrease cognitive distortions and foster positive sustainable behavioral change. Haluzan (2012) self-loathing and shame can be difficult to put into words; therefore, imagery can give form to feelings that are difficult to describe with verbal language. The use of art therapy can facilitate the expression of suppressed or discounted feelings, bringing this material to the recovering individual's awareness. This process can be pivotal in overcoming denial while encouraging individuals to become more receptive to support that is offered to them (Holt & Kaiser, 2009).

Art therapy sessions can inspire hope for individuals moving through Twelve-Step and Stages of Change Recovery Programs (Horay, 2006). The author notes there are tangible benefits to weaving art therapy and traditional motivational interviewing techniques into the fabric of these recovery models.

Gender-specific art therapy for women in forensic settings

Gussak (2004; 2009) conducted studies that found female detainees had a proclivity towards anxiety and depression due to their high prevalence of interpersonal trauma and substance abuse. Bell and Robbins (2007) used controlled art production trials to collect data supporting the hypothesis that art expression effectively reduced depression and improved the mood of participants. Day and Ororato (1997) underscore the effectiveness of art therapy groups with female inmates working towards recovery from traumatic experiences. Non-threatening, affirming and inclusive modalities, such as art therapy, can help them begin to address their traumatic histories and create a sense of belonging.

Gussak (2009) found that females' responsiveness to art therapy was significantly higher than the sample of their male counterparts. The author attributes the receptiveness to their perception of the modality being empowering as opposed to confrontational, in addition to the supportive culture evolved among group members.

Post-Traumatic Growth

Mazor, Gelkopf, & Roe (2018) found that there are significant levels of comorbidity of SMI and Posttraumatic Stress symptoms. Greater than 50% of individuals with SMI experience a traumatic incident. Mazor, Gelkopf, Muesser & Roe (2016) report upwards of 98% of individuals with psychosis have been exposed to trauma; however, limited research has been conducted on this population. Across the world post traumatic growth (PTG) is viewed as a positive psychological change that occurs after a traumatic event occurs (Thege, Kovacs & Balog, 2014).

The intense psychological struggle from a traumatic event may consist of unmistakable mental effects on a trauma survivor, despite that, ironically the struggle may bring about exceedingly meaningful insights (Calhoun & Tedeschi, 2004). Mazor et. al (2016) further explains PTG is demonstrated after the processing of thoughts, behaviors, and emotions. It is the lessening of emotional anguish and often expressed in a change of behaviors and outlook on life and past experiences. The Psychiatric Times (2004) conveyed historically human suffering was viewed as the gateway to the possibilities of something good, nevertheless there should not be the expectation of good to take place. Mazor, Gelkopf & Roe (2018) stated individuals with SMI and psychosis have the capacity to experience PTG. It is imperative to attend to both the symptoms of PTSD and the traumatic exposure for individuals with SMI. Furthermore, traditional forms of psychiatric treatment were found to be ineffective to address their life experiences and illness.

Ramos & Leal (2013) report PTG can only be recognized by the participant and merely reinforced by the group leader. PTG is reliant upon a solid therapeutic relationship, developing safety and active listening. Limitations of talk therapy interfere with furthering trauma-focused treatment once memories stored in the body are activated, leading a person to become stuck (Crenshaw, 2006). Several behavioral and cognitive approaches can be implemented to aid in the processes for meaning making of the traumatic experience and promoting PTG (Ramos & Leal, 2013).

Jayawickreme & Blackie (2016) explains two processes necessary to move a person towards PTG. The use of cognitive reprocessing of the traumatic event and by meaning making of the event. Mazor et. al (2016) also identified two important factors to contribute to the development of PTG. First, coping self-efficacy has an identified correlation to PTG. It is defined as the observed aptitude to take care of one's own environmental necessities and functioning after a traumatic incident by way of cognitive processes. The second factor is the need to pursue an understanding of the world around them following a traumatic incident. It is both a way to protect and facilitate the act of meaning making (Mazor et. al, 2016). Crenshaw (2006) states the use of creative arts therapies such as dance, music and art therapy to be effective in providing an integrated treatment approach. Establishing proper social supports within the environment are crucial to bring about PTG (Jayawickreme & Blackie, 2016).

Growth and suffering are not mutually exclusive but can co-exist simultaneously. Evidence suggest that traumatic events which often lead to negative symptoms and consequences, may also lead to renewed sense of empowerment and resiliency (Lindstrom et al., 2013). One way to foster positive adaptation is through group therapy within which one can re-process adversity with others who have also been exposed to traumatic situations. Relating to others provides opportunities to recognize paradoxes and reconstruct schemas through exploration of their own trauma narratives (Calhoun & Tedeschi, 1996). The purpose of this study was to determine whether the Empowerment Guide, which encompasses music and art therapy interventions, improves post- traumatic growth for women with severe and persistent mental illness, substance abuse and trauma histories; also defined as co-occurring disorders.

Method

This study was a single-subject quasi-experimental design. Purposive sampling, a non-probability sampling method was used to gain understanding of the specific cases within their own right instead of generalizing the results to a population (Rai & Thapa, n.d.). Specifically, the homogeneous technique, allows for emphasis on individuals with specific characteristics to provide information rich cases for a comprehensive knowledge of the sample meeting the set criterion (Etikan, Musa & Alkassim, 2016; Isaac & Michael, 1995; Patton, 2002; Rai & Thapa, n.d.).

Pre and Post testing were conducted using the Post Traumatic Growth Inventory (PTGI) to determine if the Empowerment Guide, which encompasses music and art therapy interventions, improves post-traumatic growth for women with severe and persistent mental illness, substance abuse and trauma histories; also defined as co-occurring disorders.

Participants

Four out of ten eligible participants consented to participate in the study. The hospital's electronic data base was used to retrieve the participants that met the criterion set. All participants were women residing in the hospital on long-term units. The eligible participants were 90% African-American, 5% Caucasian and 5% Latino, although 100% of the women that agreed to sign the consent form were African-American. As shown in Table 1, the ages ranged from 18-70 and the number of admissions ranged from 1-13. The education levels ranged from high school to college.

The participants met the following criterion: (1) severe and persistent mental illness, (2) trauma history, (3) minimum age of 18 years old and (4) currently housed on a long-term unit to allow opportunity for more psychological stability. The 6 non-consenting women took informed consent material and after asking clarifying questions, ultimately refused to participate. Attrition factors lead to only 2 participants completing the pre-& posttest. Participant 1 attended one session and did not return due to her concerns of overlapping treatment for trauma. Participant 2 attended all sessions except for one and ultimately refused to complete the posttest.

Table 1. Demographics

PSEUDONYM	AGE	# OF ADMISSIONS	EDUCATION LEVEL	RACE/ETHNICITY	GENDER
Participant 1	18-40	1-3	Some College	African-American	Female
Participant 2	61-70	13+	Some College	African-American	Female
Participant 3	61-70	1-3	Some High School	African-American	Female
Participant 4	51-60	1-3	Some High School	African-American	Female

Procedures

This pilot study was conducted in a large public hospital for adults who are under court ordered civil and forensic commitment in the Mid-Atlantic region. Approval from the hospital's Institutional Review Board was granted before inviting participants to join the group sessions. Consultation with each participants treatment team occurred to ensure their capacity to give consent. Pre and post testing of the Post Traumatic Growth Inventory (PTGI) were administered by professionals other than the researchers to reduce opportunity for biased responses.

Participants engaged in two 60-minute sessions with a board-certified music therapist and art therapist two times per week for three of the four designated weeks. Hospital administrative issues, such as administrative closing due to weather and administrative security protocol thwarted 2 out of the 8 sessions. Although attempts were made to reschedule the sessions, the researchers found the participants copious daily schedules did not allow for make-up sessions.

The Empowerment Guide is a 4-week multivalence approach developed by the researchers for women with severe and persistent mental illness in the forensic mental health setting. It is a synthesis of relational and gender responsive theory that addresses addictions, mental health and trauma recovery, using evidence from neuroscience through art and music. The purpose of the guide is to establish safety and reality orientation, moving towards an emphasis on resilience and self-regulation. A combination of music and art therapy interventions were used for this study. Improvisation, lyric analysis, receptive and recreative techniques such as music relaxation and listening were used in conjunction with cognitive re-structuring imagery and projective drawings.

Sessions were designed to accommodate functioning level, encourage interaction and develop comfort amongst group members. Furthermore, sessions were designed to build upon one another to increase their ability to tolerate the exploration of their traumatic experiences in a safe environment. Table 2. Provides a brief description of each week's theme and interventions.

Table 2. Empowerment Guide Sessions

Week One: All about Communication	
Session 1.	Interactive card game followed by puzzle about me and selection of preferred music was used as an ice breaker to allow each member to engage in verbal disclosure and get to know one another.
Session 2.	Each participant had to select a feeling they often experience, create a corresponding image and connect the feeling to their behaviors. Improvisation was used as a tool to non-verbally express the chosen feeling as a way of releasing the built-up tension generated by the discussion.
Week Two: Grounding & Containment	
Session 3.	Lyric Analysis opens the session to elicit distressing memories, the resolve to overcome the experience and provided validation amongst each group member. A brief psycho-educational activity follows to introduce the concept of anchoring to the "here and now." Anchoring imagery was created to promote grounding in the session.
Session 4.	Lyric Analysis opens the session to discuss the inevitable hardships in life and inspire their ability to decide how to deal with them. Next, members were asked to create a graphic image of an intense feeling using shapes and colors and subsequently created a container to store those emotions.
Week Three: Life in Review	
Session 5.	Re-Creation (Singing a song selected by MT) was used to promote reflection of the totality of their life experiences. Then comfort boxes were created and filled with imagery that provided reassurance and hope.
Session 6.	Re-creation (Singing a song selected by MT) was used to explore identity and provide opportunity to define who they are and validate personal strengths, followed by the creation of an image to symbolize how they choose to define themselves.

Empowerment Guide Model

The Empowerment Guide is one component of a holistic treatment program that addresses the complex and varied mental health, addictions, and trauma-focused needs of women in the forensic hospital setting. The guide is a synthesis of relational and gender-specific theory using music and art as modalities for treatment. It combines two highly effective approaches to trauma-focused -treatment: creative arts therapies and cognitive-behavioral therapy. Opportunities for expression, containment and education are provided in equal measure. Learning concepts are simplified to accommodate the needs of SMI clients, many of whom have cognitive impairments (Gold & Harvey, 1993). Problems with attention, concentration as well as limitations in abstract reasoning are also given consideration. The acquisition of new skills and new ways of thinking optimizes healing and minimize the impact of stress.

The interplay between art, music and cognitive reprocessing can create a therapeutic space for self-awareness and expression. Opening sessions to two or more creative arts modalities, enables group members to express themselves in a manner, and at a rate, which is most appropriate for them (Knill et. al, 1995). The creative nature of art and music are leveraged to facilitate exploration of painful memories to transform their meaning. Rogers (1993) suggests that using more than one art-based intervention can build off one another to create deeper awareness. Art and musical expression also provide a springboard for discussion where cognitive restructuring strategies are used to help reshape distorted beliefs, thoughts and expectations (Mauritz et al., 2013).

Cognitive techniques such as re-processing and reframing can be employed to create personal narratives in ways that are creative and unique. Through the process of re-storying, one can redefine their identity from trauma survivor to someone who was faced with adversity and chose to become a stronger person (Johnson, 2009).

Effects of traumatic exposure can have a disruptive impact in many areas of mental health and biopsychosocial functioning. The three-phase model proposed by Herman (1997) provides a simple yet holistic framework for the

Empowerment Guide. Its clinically sequenced stages create the safety necessary to achieve symptom reduction and stabilization. Herman (1997) views the group format as the driving force for recovery. A sense of belonging can buffer pain by providing compassion and instilling the knowledge that the individual is not alone.

Group members process situations that occur within the setting, generalizing what is learned in sessions to their day-to-day lives. It is the connection with one another, and they bear witness to each other's pain, while encouraging peers to challenge negative core beliefs. Rogers (1993) applies person-centered concepts to the creative arts therapies; recognizing all people have the capacity for self-direction, growth and the realization of their full potential.

Measures

The group participants perceptions of positive change prior to, and following, their participation in the Empowerment Guide groups were assessed using The Post-traumatic Growth Inventory (PTGI). It is an instrument designed to assess positive outcomes reported by individuals who have experienced traumatic events (Tedeschi & Calhoun, 1996). The PTGI is a 21 item-scale self-report inventory, which evaluates whether positive changes have occurred in response to significant life adversities.

The scoring occurs on a 6-point Likert scale, with 0 equaling *"I did not experience this change as a result of my crisis"* to 5 *equaling "I experienced this change to a very great extent as a result of my crisis"* (Mazor, Gelkopf, Mueser & Roe, 2016). The PTGI has exceptional internal consistency (α=.90) and maintains acceptable test-re-test reliability (*r*=.71). Supported evidence of validity (*r*=.69) of the PTGI's responses tended to be verified by those close to the persons reporting growth (Cann et. al, 2010)

Results

A paired samples t-test was used to examine any potential differences between the pre-test and post-test of the five factors of the PTGI administered to the participants. The sample size *N=2*. Factor I Relating to others; the difference between pre-test (*M* = 3.43, *SD* = 1.41) and post-test (M = 3.78, SD = .11) on the relating to others scale was not statistically significant, t(1) = .32, p = .80. Factor II New Possibilities; the difference between pre-test (M = 3.5,

SD = 1.27) and post-test (M = 4, SD = 0) onthe new possibilities scale was not statistically significant, t(1) = .55, p = .67. Factor III is Personal Strength. The difference between pre-test (M = 3.37, SD = 1.59) and post-test (M = 3.62, SD = .17) on the personal strength scale was not statistically significant, t(1) = .2, p = .87. Factor IV is Spiritual Change. The difference between pre-test (M = 3.75, SD = 1.76) and post-test (M = 4.5, SD = .70) on the spiritual change scale was not statistically significant, t(1) = .42, p = .74. Factor V is Appreciation. The difference between pre-test (M = 3, SD = 2.82) and post-test (M = 3.17, SD = .70) on the appreciation scale was not statistically significant, t(1) = .11, p = .92. The null hypothesis stands as in all five factors there was no statistically significant difference. Table 3 displays the Pre and Post-test of the Five Factors in the PTGI.

Table 3. Mean and SD of the Five Factors

Outcome	Pre-test		Post-test				
	M	SD	M	SD	t	df	p-value
Factor I Relating to others	3.43	1.41	3.78	.11	.32	1	.80
Factor II New Possibilities	3.5	1.27	4	0	.55	1	.67
Factor III Personal Strength	3.37	1.59	3.62	.17	.2	1	.87
Factor IV Spiritual Change	3.75	1.76	4.5	.70	.42	1	.74
Factor V Appreciation of Life	3	2.82	3.17	.70	.11	1	.92

Limitations

Due to the unique, but common population in this study yielded a small sample size. Factors such as scheduling, potential participants unwillingness to sign consent forms, concerns with court cases, and unfamiliarity with the researchers hindered recruitment processes. Additionally, the nature of the severe mental illness also contributed to the attrition of those initially agreeing and signing consent forms for participation. There were also limitations of the design approach and the unique qualities of this population restricted the researcher's ability to determine whether the empowerment guide improved post-traumatic growth in women with co-occurring disorders from a qualitative perspective.

Discussion

Despite the growing body of research on trauma exposure and PTSD, literature on trauma treatment with SMI populations is sparse. A lack of recognition of trauma-related symptoms in SMI individuals is one factor (Mauritz et al., 2013). Making a distinction between dissociative and psychotic episodes is difficult due to the overlapping nature of their symptoms. Additionally, there is concern regarding the vulnerability of SMI individuals to stress (Mauritz et al., 2013). Caution about furthering impairment may divert clinical focus away from treating trauma-related symptoms (Kilcommon & Morrison, 2005). Some assert that such concerns are not substantiated by research (Griffin, Resick, Waldrop & Mechanic, 2003).

Others assert that SMI individuals are more sensitive to stress than the general psychiatric population; therefore, stress associated with recall of distressing memories should be mitigated (Butzlaff & Holley, 1998). There is remarkably little data available to serve as a guide when providing clinically sound trauma intervention to individuals living with SMI (Mueser, Stanley & Janjowski, 2004).

Creative arts therapies can be used as a medium to explore meaning, bring change and dive into the unconsciousness (Megranahan & Lynskey, 2018). The use of illicit drugs may often become a means by which women deal with domestic violence and sexual trauma.

Furthermore, incarcerated women are more likely than men to suffer from mental illness.

Although it is known of the need for more effective trauma-focused interventions for the SMI population, development of normalized treatment programs has yet to be empirically substantiated (Mueser et. al, 2004). Even though women are housed in facilities that were built for men, women require innovative programs that are responsive to their gender-specific needs (Covington, 2007). There continues to be a need to develop more effective trauma focused treatment for women with SMI, addiction and trauma histories. A decade ago, the facility implemented a person-centered recovery model to promote autonomy and resiliency in individuals in its care, many of whom were and continue to be, living with SMI. In response, the Creative Arts Therapy Department launched several initiatives to help SMI individuals recapture a cohesive sense of self, bridge them to the community at large and develop trauma-focused treatment using the arts.

Arts-based treatment interventions are an integral part of the daily lives of the consumers. A safe environment must be created, and the therapist must have the strength to support the trauma survivor participating in the trauma focused groups (Sutton, 2002).

Calhoun (2004) expresses the importance of creating a socially supportive environment to promote PTG.

Empowerment Guide accounts for the sensitive and unique needs of this population and was designed to reduce the intensity of stress when engaging in the trauma-focused treatment. While results from the paired sample t test were not statistically significant, having a qualitative component would provide additional context and better inform the effectiveness of the Empowerment Guide.

The nature of SMI also hinders participants agreeing and committing to the research process; which makes it difficult to maintain a reasonable sample size. The results of this study inform music and art therapy clinicians of the need for a mixed method design to better inform gender specific treatment. Although, there were confines to the design approach, participants expressed the skills that were gained, self-awareness to enhance autonomy and self-regulation, as well as an awareness of any growth after traumatic experiences.

Future Research

While the findings from this pilot study contribute to the body of existing research on women with co-occurring disorders in the forensic psychiatric hospital setting, the following should be taken into consideration for future research.

1. A larger sample size should be considered to further expand the findings of this study and to increase the likelihood of statistical significance.
2. A similar study should be conducted repeating the modules from the Empowerment Guide with the same sample.
3. A longitudinal study of women with co-occurring disorders should be conducted for those participating in the Empowerment Guide to understand the treatment process more over a period of time and the potential for post traumatic growth.
4. A comparison analysis of women with co-occurring disorders participating in "treatment as usual" versus those participating in the Empowerment Guide sessions.

References

Ahonen, H. (2016). Adult Trauma Work in Music Therapy. In S. Hallam, I. Cross, & M. Thaut (Eds.), The Oxford Handbook of Music Psychology (2 ed.). (pp.1-29). doi: 10.1093/oxfordhb/9780199639755.013.47

Alemagno, S.A. (2001). Women in jail: Is substance abuse treatment enough? *Journal of Public Health.* 91, 796-800.

Amir, D. (2004). Giving Trauma a Voice: The Role of Improvisational Music Therapy in Exposing, Dealing with and Healing a Traumatic Experience of Sexual Abuse. *Music Therapy Perspectives.* 22(2), 96.

Barnes, M., Cohen, B. & Rankin, A. (1995) Managing traumatic stress through art. Luthurville, MD: Sidran Press.

Bell, C. E., & Robbins, S. J. (2007). Effects of art production on negative moods: A randomized, controlled trial. *Art Therapy: The American Journal of the Art Therapy Association.* 24(4), 71-75.

Butzlaff,, R. L., & Hooley, J. M. (1998). Expressed emotion and psychiatric relapse. *Archives of General Psychiatry*. 55, 547-552.

Calhoun, L. G., & Tedeschi, R. G. (2004). The Foundations of Posttraumatic Growth: New Considerations. *Psychological Inquiry*. 15(1), 93-102.

Calhoun, L. & Tedeschi, R. (1996). The post traumatic growth inventory. Journal of Traumatic Stress. 9(3), 455-469.

Cann, A., Calhoun, L. G., Tedeschi, R. G., Taku, K., Vishnevsky, T., Triplett, K. N., & Danhauer, S. C. (2010). A short form of the Posttraumatic Growth Inventory. *Anxiety, Stress, & Coping.* 23(2), 127-137.

Carr, C., d'Ardenne, P., Sloboda, A., Scott, C., Wang, D. & Priebe, S. (2012). Group music therapy for patients with persistent post-traumatic stress disorder-an exploratory randomized controlled trial with mixed methods evaluation. *Psychology and Psychotherapy: Theory, Research and Practice.* 85, 179-202.

Crenshaw, D. (2006). Neuroscience and Trauma Treatment: Implications for Creative Arts Therapists. In L. Carey (Ed.), Expressive and creative arts method for trauma survivors. (pp. 21-29). Retrieved from http://ebookcentral.proquest.com

Creswell, L.M. (2014). A critical black feminist ethnography of treatment for women with co-occurring disorders in the psychiatric hospital. *The Journal of Behavioral Health Services & Research,* 41(2), 167–184.

Creswell, L. (2016). Review of Literature of Women with Co-occurring Disorders through the Feminist and Black Feminist Perspective. Journal of Addiction Research & Therapy. 7, 267. doi:10.4172/2155-6105.1000267

Cohen, L.R. & Hien, D.A. (2006). Treatment outcomes for women with substance abuse and PTSD who have experienced complex trauma. *Psychiatry Services*, 57, 100-106.

Covington, S. (2007) Women in the criminal justice system. *Women's Health Issues.* 17, 180-182.

Covington, S. S., & Surrey, J. L. (2000). The Relational Model of Women's Psychological Development: Implications for Substance Abuse. Ins S. Wilsnack & R.Wilsnack (Eds.), *Gender and Alcohol: Individual and Social Perspectives* (pp.335-351). New Brunswick, NJ: Rutgers Center of Alcohol Studies.

Dass-Brailsford, P., & Safilian, C. M. (2017). Integrate Approaches to Treating Psychological and Substance Abuse in Women: An Update. *Trauma Acute Care.* 2, 41.

Day, E. & Ororato, (1997). Surviving one's sentence: Art therapy with incarcerated trauma survivors. In D. Gussak, & E. Virshup (Eds.) Drawing time: Art therapy in prisons and other correctional facilities. (pp. 127-152). Chicago, IL: Magnolia Street Publishers.

Donovan, J., Molina, B., & Kelly, T. (2009). Alcohol outcomes expectancies as socially sharedand socialized beliefs. *Psychology of Addictive Behaviors.* 23, 248-259.

Etikan, I., Musa, S. A., & Alkassim, R. S. (2016). Comparison of Convenience Sampling and Purposive Sampling. *American Journal of Theoretical and Applied Statistics.* 5(1), 1-4.

Gardstrom, S., Klemm, A., & Murphy, K. (2017). Women's Perceptions of the Usefulness of Group Music Therapy in Addictions Recovery. *Nordic Journal of Music Therapy.* 26(4), 338-358.

Gold, J., & Harvey, P. (1993). Cognitive deficits in schizophrenia. *Psychiatric Clinics of North America.* 16(2), 295-312.

Gold, C., Mossler, K., Grocke, D., Heldal, T. O., Tjemsland, L., Aarre, T., Aaro, L. E., Rittmannsberger, H., Stige, B., Assmus, J. & Rolvsjord, R. (2013). Individual Music Therapy for Mental Health Care Clients with Low Therapy Motivation: Multicentre Randomised Controlled Trial. *Psychotherapy and Psychosomatic*. 82, 319-331.

Gold, C., Rolvsjord, R., Aaro, L. E., Aarre, T., Tjemsland, L. & Stige, B. (2005). Resource-oriented music therapy for psychiatric patients with low therapy motivation: Protocol for a randomized controlled trial. *BMC Psychiatry*. 5, 39. doi: 10.1186/1471-244X/5/39

Griffin, M. G., Resick, P. A., Waldrop, A. E., & Mechanic, M. B. (2003). Participation in trauma research: Is there evidence of harm? *Journal of Traumatic Stress*. 16, 221-227.

Grocke, D., Bloch, S., Castle, D., Thompson, G., Newton, R., Stewart, S. & Gold, C. (2014). Group music therapy for severe mental illness: a randomized embedded-experimental mixed methods study. *Acta Psychiatrica Scandinavica*. 130, 144-153. doi: 10.1111/acps.12224

Gunn, A. J., Sacks, T. K., & Jemal, A. (2016). "That's not me anymore": Resistance strategies for managing intersectional stigmas for women with substance use and incarceration histories. *Qualitative Social Work*. 0(00), 1-19. doi: 10.1177/1473325016680282

Gussak, D. (2004) A pilot research study on the efficacy of art therapy with prison inmates. *The Arts in Psychotherapy.* 31(4), 245-259.

Gussak, D. (2009) The effects of art therapy on male and female inmates: Advancing the research base. *The Art in Psychotherapy.* 36(4) 5-12.

Gussak, D. E. & Ploumis-Devick, E. (2004). Creating Wellness in Correctional Populations Through the Arts: An Interdisciplinary Model. *Visual Arts Research* 30(1), 35.43.

Haluzan, M. (2012). Art therapy in the treatment of alcoholics. *Alcoholism.* 48(2), 99-105.

Herman, J. (1992). Trauma and Recovery: The aftermath of violence from domestic abuse to political terror. New York, NY: Basic Books.

Hodges, D. A. (2016). The Neuroaestheticsof Music. In S. Hallam, I. Cross, & M. Thaut (Eds.), The Oxford Handbook of Music Psychology (2 ed.). (pp.1-22). doi: 10.1093/oxfordhb/9780198722946.013.20

Holt, E.C., & Kaiser, D.H. (2009). The First Step Series: Art therapy for early substance abuse treatment. *The Arts in Psychotherapy*, 36(4), 245-250.

Horay, B.J. (2006). Moving towards gray: Art therapy and ambivalence in substance abuse treatment. *Art Therapy Today.* 23(1), 12- 22.

Isaac, S., & Michael, W. B. (1995). *Handbook In Research And Evaluation (3rd ed.). San Diego, California: EdITS.*

Jayawickreme, E., & Blackie, L. E. (2016). *Exploring the psychological benefits of hardship*. Berlin: Springer. doi:10.1007/978-3-319-47989-7_2

Johnson, D.R. (2007). Examining underlying paradigms in the creative arts therapies and trauma. *The Arts in Psychotherapy*. 36, 114-120.

Kilcommon, A. M. & Morrison, A.P. (2005) Relationships between trauma and psychosis: An exploration of cognitive and dissociative factors. *Acta Psychiatrica Scandinavica*. 112, 351-359.

Knill,P., Barba, H. & Fuch, M. (1995). Minstrels of the Soul: Intermodal expressive therapy. Toronto: EGS Press.

Koelsch, S. (2015, March). Music-evoked emotions: principles, brain correlates, and implications for therapy. *Annals of The New York Academy of Sciences*. 1337, 193-201. doi: 10.1111/nyas.12684.

Lindstrom, C. M., Cann, A., Calhoun, L. G., Tedeschi, R. G. (2013). The relationship core belief, rumination, disclosure and sociocultural elements of posttraumatic stress. *Psychological Trauma*. 5(1), 50-55.

Malchiodi, C. A. (2001). *Using drawing* as inter-vention with *traumatized children*. *Trauma* and *loss*: *Research* and *interventions*. *1*(*1*), *21-27*.

Mauritz, M. W., Goossens, P. J. J., Draijer, N. & van Achterbert, T. (2013). Prevalence of interpersonal trauma exposure and trauma-related disorders in severe mental illness. *European Journal Psychotraumatology*. 4, 1-15.

Mazor, Y., Gelkopf, M., Mueser, K. T. & Roe, D. (2016). Posttraumatic Growth in Psychosis. *Frontiers in Psychiatry*. 7(202), 1-11.

Mazor, Y., Gelkopf, M., & Roe, D. (2018). Posttraumatic growth among people with serious mental illness, psychosis and posttraumatic stress symptoms. *Comprehensive Psychiatry*. 81, 1-9.

McCaffrey, T., Edwards, J. & Fannon, D. (2011). Is there a role for music therapy in the recovery approach in mental health? *The Arts in Psychotherapy* 38, 185–189.

Meekums, D. (1999). A creative model for recovery from child sexual abuse trauma. *The Arts in Psychotherapy*. 26, 247-259.

Megranahan, K. M., & Lynskey, M. T. (2018). Do creative arts therapies reduce substance misuse? A systematic review. *The Arts in Psychotherapy*,57, 50-58.

Moore, R. (1983). Art therapy with substance abusers: A review of literature. *The Arts and Psychotherapy*. 10, 251-260.

Mueser, K. T., Rosenberg, S. D., Hamblen, J. L., & Descamps, M. (2004). A Cognitive–Behavioral Treatment for Posttraumatic Stress Disorder in Severe Mental Illness. *American Journal of Psychiatric Rehabilitation*, *7*, 107-146. doi:10.1080/15487760490476183

NIDA. (2018, August 10). Substance Use in Women. Retrieved from https://www.drugabuse.gov/publications/drugfacts/substance-use-in-women on 2019, May 8

Patton, M. Q. (2002). *Qualitative Research & Evaluation Methods (3rd ed.)*. Thousand Oaks: Sage Publications, Inc.

Pomeroy E. C., Kiam, R. &Abel, E. (1998) Meeting the mental health needs of incarcerated women. *Health &Social Work, 23*, 71-76.

Rai, N., & Thapa, B. (n.d.). A Study on Purposive Sampling Method in Research. Retrieved from https://www.academia.edu/28087388/A_STUDY_ON_PURPOSIVE_SAMPLING_METHOD_IN_RESEARCH

Ramos, C., & Leal, I. (2013). Posttraumatic Growth in the Aftermath of Trauma: A Literature Review About Related Factors and Application Contexts. *Psychology, Community & Health*. 2(1), 43-54. doi: 10.596/pch. v2i1.39

Rogers, N. (1993) The Creative Connection: Expressive Therapy as Healing. Palo Alto, CA: Science & Behavior Books.

Rolvsjord, R. (2015). What clients do to make music therapy work: A qualitative multiple case study in adult mental health care. *Nordic Journal of Music Therapy*. 24(4), 296-321.

Rothchild, B. (2000). *The body remembers: The psycho-physiology of trauma and trauma treatment*. New York: Norton.

Sanders, M. (n.d.). MISA Treatment for Women: A Focus on Trauma. Retrieved January 7, 2019, from https://irp-cdn.multiscreensite.com/38c63840/files/uploaded/Misa Treatment For Women A Focus On Trauma.pdf

Sareen, J. (2014). Posttraumatic Stress Disorder in Adults: Impact, Comorbidity, Risk Factors, and Treatment. *Canadian Journal of Psychiatry,* 59(9), 460-467.pdf

Seigal D. J. (1999). *The developing mind: Towards a neurobiology of interpersonal experience.* New York: Guildford Press.

Steele, W., & Raider, M. (2001.) Structured sensory intervention with traumatized children, adolescents and parents. *Trauma and Loss Research and Intervention*. 1(1), 8-20.

Stein, S. M. & Friedman, P. D. (2001). Generalist physicians and addictions care. *Journal of American Medical Association*. 286, 1764-1765.

Substance Abuse and Mental Health Services Administration (2014). SAMHSA's Concept of Trauma and Guidance for a Trauma-Informed Approach. HHS Publication No. (SMA) 14-4884. Rockville, MD: Substance Abuse and Mental Health Services Administration.

Sutton, J. P. (2002). Trauma: Trauma in context. In J. P. Sutton (Ed.), *Music, Music Therapy and Trauma* (pp.21-39). London: Jessica Kingsley Publishers.

Swallow, M. (2002). Neurology: The Brain-Its Music and its Emotion: The Neurology of Trauma. In J. P. Sutton (Ed.), *Music, Music Therapy and Trauma* (pp.41-53). London: Jessica Kingsley Publishers.

Talwar, S. (2007). Accessing traumatic memory through artmaking: An art therapy protocol. *The Arts in Psychotherapy.* 34, 22-35.

Tedeschi, R. & Calhoun, L. (1996). The Posttraumatic Growth Inventory: Measuring the positive legacy of trauma. *Journal of Traumatic Stress,* 9(3), 455-471.

Tedeschi, R. G., & Calhoun, L. (2004, April 01). Posttraumatic Growth: A New Perspective on Psychotraumatology. *Psychiatric Times.* 21(4). Retrieved from: https://www.psychiatrictimes.com/ptsd/posttraumatic-growth-new-perspective-psychotraumatology

Thege, B. K., Kovacs, E., & Balog, P. (2014). A bifactor model of Posttraumatic Growth Inventory. *Health Psychology & Behavioural Medicine*. 2(1), 529-540.

van den Berg, D. P. G., de Bont, P.A.J. M.,vander Vleugel, B. M., de Roos, C., de Jongh, A., van Minnen, A., & van der Gaag, M. (2016). Trauma- Focucsed Treatment in PTSD Patients With Psychosis: Symptom Exacerbation, Adverse Events, and Revictimization. *Schizophrenia Bulletin,* 42(3), 693-702. doi: 10.1093/schbul/sbv172

Vogt, D. (2018, September 24). Research on Women, Trauma and PTSD. *Psychological Bulletin.* 133, 183 204. doi: 10.1037/0033-2909.133.2.183

Williams, E., Dingle, G. A., Jetten, J. & Rowan, C. (2019). Identification with arts-based groups improves mental wellbeing in adults with chronic mental health conditions. *Journal of Applied Social Psychology.* 49, 15-26.

Healing Frequencies 174hz-963hz

Decoded from the Bible

Carla Rose Kelly & Julia MacKinley

Abstract

Secret healing frequencies that were hidden in the Book of Numbers of the Bible for over 2000 years have recently been decoded by Dr. Puleo. There is also a Biblical story from the Book of Samuel describing David playing the harp to heal King Saul's madness, but the exact frequencies and number of strings were unknown. Are these newly discovered healing numbers, the frequencies on the miraculous harp that David used to heal King Saul's madness? There is no evidence to know for sure.

Modern day, patented musical instruments have been made with the healing frequencies decoded from the Bible: 174hz, 285hz, 396hz, 417hz, 528hz, 639hz, 741hz, 852hz, 963hz. They have been used as music for medicine, in a clinical setting in London Ontario, Canada for the past 10 years, treating clients with depression, anxiety and pain. The research on this topic has shown it is effective in alleviating symptoms of pain, depression and anxiety and could be used in a wide array of disorders.

Introduction

In the King James Bible, Old Testament, Book of Samuel XVI: 14-23, a story is told of how King Saul calls for David to play the harp to cure his ills. David came to King Saul, and cured his madness. Music was the medicine at this time, and the harpist was the "doctor" by today's standards. In this biblical story, we are not given any information to indicate the frequencies, or number of strings used on this magical harp that David used to cure King Saul illness. Also, there aren't any records to indicate what sorts of songs, or tones they were playing to cure the ills. What we know from our earliest written records is outlined below. However, there is no evidence to suggest this is how David, and many other music healers of their time, played their music.

Our earliest written records in the history of music only indicate that music was monophonic, that is melody without harmony or counter-point. Secondly, we know from our earliest records, that the music was all improvised. The performer was the composer. This does not mean that what he did was completely spontaneous and unprepared. The musician/composer had to keep within the universally accepted rules governing the forms and styles of music suitable for particular occasions. It is probable that he incorporated in his performance, certain traditional musical formulas. He wasn't playing something from a score, and consequently no two performances of the "same" piece, were exactly alike. This practice prevailed into our Western music up to the eighth century A.D. Thirdly, we know that this music was always associated with words or dancing or both. The singers used the melody and rhythm of poetry and performed it with movements of prescribed dance patterns. Fourth, we have information about these early Greek theories, but not about the music itself. The Greeks had doctrines of the nature of music, its place in the cosmos, its effects, and its proper uses in human society. They achieved insights and formulated principles of using Music as Medicine, many of which have not been superseded to this day. Pythagoras was one of these Greek philosophers that understood how music could be used for healing and to reach union with Divine. Part of the teachings was

Numerology which is the spiritual study of numbers, Pythagoras taught in 500BC. He understood that music and math were not separate; as the understanding of numbers was thought to be the key to the understanding of the whole spiritual and physical universe, so the system of musical sounds and rhythms, being ordered by numbers, was conceived as exemplifying the harmony of the cosmos and corresponding to it. There was a belief that there was a power in music akin to the power of words for influencing human thought and action. It was felt that an artist is under obligation to exercise this power with due regard for its effect on others. This was one of the most important aspects of Greek thought about music. Music was a microcosm, a system of sound and rhythm ruled by the same mathematical laws that operate in the whole visible and invisible creation. Music was not only a passive image of the orderly system of the universe; it was also a force that could affect the universe. Hence, we have the attribution of miracles to the mythological Gods of Music & Medicine, like Apollo. In classical Greek mythology, there were nine sister goddesses who presided over the arts and science. Music was thought of as something basic to activities concerned with the pursuit of truth (Grout et al., 2010).

Today, people are trying to recreate a healing harp like David may have used to heal King Saul's madness. Modern day speculations to the number of strings and exact frequencies, remain purely speculative. This is why our modern-day Music & Medicine research is so important in helping us understand the implications of rhythm, melody, tone, etc, and its effect on the health of an individual. It is only when we unlock all these secrets that our ancestors understood as commonplace, that we can begin to understand how illness can be treated, and perhaps prevented by music.

Healing Numbers Decoded in Bible

Although the number of strings and exact frequencies of the healing harp that David used were not directly indicated in the bible, hidden information was recently decoded, relating to Numerology, that may provide us with clues. However, there is no conclusive evidence that the following information relates to these mysterious frequencies.

Within modern Torahs, and the King James Bible, Book of Numbers, Chapter 7, verses 12-83, repeating verses camouflaged six numbers, according to Dr. Joseph Puleo of Idaho, USA. These numbers he decoded were the following:

1. 396
2. 417
3. 528
4. 639
5. 741
6. 852

He used the Pathagorean skein to decipher each verse number down to a single digit, by adding up the numbers within it. For example, in the Pathagorean skein, if you had the number 396, you add up the 3+9+6=18. Then you add 1+8=9, reducing it down to a single digit. Dr. Puleo was under the spiritual guidance of "Jeshua", whom he prayed to for understanding of the repeating verses. Dr. Puleo consulted with Levi priests who translated the original Torah into the Greek Septuagint, whom thought these six frequencies possessed extraordinary spiritual powers. These Levi priests also felt the frequencies were linked to the hymn of St. John the Baptist. They indicated that MI represented "Miracles". Dr. Puleo concluded the 6 uncovered numbers were related to the Solfeggio Frequencies. Ut was first used before we changed it to Do, hence our solfeggio system. The following shows the corresponding solfeggio sound that Dr. Puleo put with the frequency, and the right column indicates the single digit it corresponds to according to the Pythagorean skein. The solfeggio will be addressed later in this paper.

1. Ut = 396 = 9
2. Re = 417 = 3
3. Mi = 528 = 6
4. Fa = 639 = 9
5. Sol = 741 = 3
6. La = 852 = 6

(Horowitz et al., 1999)

From these six numbers, using the Pythagorean skein and teachings, Dr. Puleo figured out the missing three numbers: 174, 285, and 936. The six were not complete, because in Numerology, there are 9 root numbers which correspond to the understanding of the whole universe and its creation. All other numbers are reduced to a single digit between one and nine. Here is the complete list of hidden numbers decoded by Dr. Puleo with their single digit reduction obtained by adding up the numbers within it:

174 = 1+7+4 = 12 = 1+2 = 3

285 = 2+8+5 = 15 = 1+5 = 6

396 = 3+9+6 = 18 = 1+8 = 9

417 = 4+1+7 = 12 = 1+2 = 3

528 = 5+2+8 = 15 = 1+5 = 6

639 = 6+2+9 = 18 = 1+8 = 9

741 = 7+4+1 = 12 = 1+2 = 3

852= 8+5+2 = 15 = 1+5 = 6

963= 9+6+3 = 18 = 1+8 = 9

Let's examine the nine numbers down the left column, 174-936. If you were to separate each of the nine numbers into a single column, you will notice something interesting. You have a complete set of one through nine on the left column. The middle column starts with a seven, continues through eight, nine, then starts over at one through six, again, another complete set of nine numbers. The right column begins with four through nine, then begins with 1 finishing with three, again, another complete set of nine numbers. All 3 columns are a complete set of one through nine.

1	7	4
2	8	5
3	9	6
4	1	7
5	2	8
6	3	9
7	4	1
8	5	2
9	6	3

If you were to continue this number pattern after 963, the next number would be 174, which is the beginning number. It appears to be a complete set of numbers that continually repeats itself, and no other numbers are necessary.

You may have also noticed that when each of the nine numbers was reduced to a single digit, a repeating cycle of three, six, nine occurs, which is shown three times, as indicated in the right column. Symbology and spirituality recognize the three as the Holy Trinity, the six is the Holy Trinity twice, and the nine, is the Holy Trinity three times. The Holy Trinity is equated in religions around the world to Creation.

Solfeggio?

Earlier in this paper, it was indicated that the Levi priests advised Dr. Puleo that the six numbers were related to the Hymn of St. John the Baptist and the solfeggio system. It is unknown whether these priests had a musical background or knew about the

Ut Queant Laxis (Hymn to St. John the Baptist)

Guido of Arezzo
(circa 991-1033)

Translation:

So that your servants may, with loosened voices, resound the wonders of your deeds, clean the guilt from our stained lips, O Saint John.

history of music. According to the Royal Conservatory's accepted text book, History of Music written by Donald Jay Grout, an eleventh-century monk, Guido of Arezzo, perfected a method for the teaching of sight singing. It had as its basis, the memorizing of six tones in the pattern C-D-E-F-G-A; in this pattern a semitone falls between the third and fourth steps and all other steps are whole tones. Guido pointed out, as an aid to memorizing the pattern, that in a familiar hymn, *Ut queant laxis*, (Hymn to St. John the Baptist), each of the six phrases began with one of the notes of the pattern in regular ascending order. The first phrase on C, the second on D, and so on. The initial syllables of the words of these six phrases became the names of the notes: ut, re, mi, fa, sol, la. We still teach them this way today, except we have changed ut to do, and have added ti and la. The semitone, it will be noted, is always the interval mi-fa. This is our solfeggio system, the movable do system, the basis for our diatonic scale we use today. When one looks at this hymn below, you can clearly see Guido d'Arrezzo's use of the treble clef, and each of the corresponding six notes: ut-C, re-D, mi-E, fa-F, sol-G, la-A.

The following chart shows our modern-day frequencies as they are related to the original solfeggio system and corresponding musical letter names. The mathematical intervals between each note in the diatonic scale is clearly not the same as the numbers decoded from the Bible. It is also very clear that they are not the same frequencies as well. We see the high octave of C as 528, but none of the other frequencies are identical. Calling the decoded numbers, Solfeggio, may have been misunderstood by the Levi priests, had they known the history of our solfeggio system of music, and its origins to our diatonic scale we use today. However, the world has accepted the newly decoded frequencies by Dr. Puleo by the name of Solfeggio, so that is their title now. It may be confusing to music teachers who teach solfeggio as a basis to the diatonic scale, and now it serves also as the title of a healing scale. We will need to know the difference.

Diatonic Scale

Letter	Note	Frequency Hz
C	Do	264
D	Re	297
E	Mi	330
F	Fa	352
G	Sol	396
A	La	440
B	Si	495
C'	do	528

Definitions of the Solfeggio Healing Frequencies

There is much information on the internet about the definitions of each frequency and what each claim to do. It was stated that Dr. Puleo along with other scientists did extensive research on the frequencies. However, this research could not be found, so it was not available to be included in this paper. It should be noted that the comments from persons listening to the frequencies on the internet, are very positive and supportive of the claims for each frequency. The following information describes the original six Solfeggio frequencies and what each claim to do.

The main six Solfeggio frequencies are:

- **396 Hz –** Liberating Guilt and Fear
- **417 Hz –** Undoing Situations and Facilitating Change
- **528 Hz –** Transformation and Miracles (DNA Repair)
- **639 Hz –** Connecting/Relationships
- **741 Hz –** Expression/Solutions
- **852 Hz –** Returning to Spiritual Order

According to Dr. Leonard Horowitz who worked very closely with Dr. Puleo, 528 Hertz is a frequency that is central to the "musical mathematical matrix of creation." More than any sound previously discovered, the "LOVE frequency" resonates at the heart of everything. It connects your heart, your spiritual essence, to the spiraling reality of heaven and earth. Math scientist Victor Showell describes 528 Hz as fundamental to the ancient Pi, Phi, and the Golden Mean evident throughout natural design. This is discussed in detail below. Vic Showell and John Stuart Reid (a pioneer in acoustic research and cymatic measurements) have proven that 528 is essential to the sacred geometry of circles and spirals consistent with DNA structuring and hydrosonic restructuring. They claim that listening to 528Hz promotes personal power and spiritual growth. It is also claimed that it helps us release unconscious blockages and limitations that have held us back, so that we can with lesser resistance become our authentic self and move into the light of our higher mind. (Jakowiak, 2019)

Do the Nine Solfeggio Frequencies Contain Spiritual Properties?

Yes, it would appear that the nine Solfeggio frequencies may contain spiritual powers that can heal and help humanity align with Divine. When we look at the math underlying the Solfeggio frequencies, all nine contain the harmonics of the Fibonacci Spiral which explains creation mathematically. Each number in the Fibonacci Spiral approximates PI, PHI, or the Golden Mean. The Fibonacci Spiral is a series of numbers that mathematically explain how everything within creation grows from the beginning out into infinity. The math within all things in nature is exactly the same; everything is made in the same likeness. This was determined by Leonardo Fibonacci (1175-1250 AD, Italy), one of the greatest mathematicians of all times. He spent time in nature studying how the flowers and trees grew. He compared the mathematics for plant growth with how animals multiplied if left uninterrupted by humans. Fibonacci compared this information with how the human body is made and finally to the formation of our universe. He conclusively found the same pattern existed for all things in nature, he named Fibonacci Series. The following is the Fibonacci Series:

0, 1, 1, 2, 3, 5, 8, 13, 21, 34, 55, 89, 144, 233, 377, 610...infinity

The pattern within creation is simple. The Fibonacci Spiral is created by adding the first two numbers, (0 + 1) to get the next number (1), then you add the second and third numbers (1 + 1) to get the fourth number (2). You continue adding up the next two consecutive numbers to get each subsequent number. (Fibonacci, 1202)

The Nine Solfeggio Frequencies Contain the Harmonics of Creation

The following is an outline of each of the Solfeggio frequencies and how each one contains the vibrations of these creational numbers over and over again. This presentation also demonstrates how each of the nine Solfeggio frequencies carries the same harmonies which explains why they are all in tune with each other.

When one's ears become attuned to it, you can hear the similar harmonics within the frequencies. Because they hold within them, the creative energy, this presents a very powerful creative spiraling energy, similar to that of nature itself, or God or universal powers, etc.

Again, the Fibonacci Series:

0, 1, 1, 2, 3, 5, 8, 13, 21, 34, 55, 89, 144, 233, 377, 610, ... infinity

174 is the sum of 7 sets of Fibonacci numbers:

(1+8+21+144), (1+3+5+21+144), (1+8+21+55+89), (1+3+5+8+144), (1+3+5+21+55+89), (1+3+5+9+13+55+89), (1+3+5+8+13+21+34+89)

285 is the sum of 9 sets of Fibonacci numbers:

(5+13+34+233), (2+3+13+34+233), (5+13+34+89+144), (2+3+13+34+89+144), (2+3+5+8+13+21+233), (2+3+5+8+13+21+89+144), (2+3+5+8+13+21+34+55+144)

396 is the sum of 12 sets of Fibonacci numbers:

(1+5+13+377), (1+2+3+13+377), (1+5+13+144+233), (1+2+3+13+144+233), (1+5+13+55+89+233), (1+2+3+13+55+89+233),

(1+5+13+21+34+89+233),

(1+2+3+5+8+55+89+233),

(1+2+3+13+21+34+233),

(1+2+3+5+8+21+34+89+233)

417 is the sum of 15 sets of Fibonacci numbers:

(2+3+5+8+34+233), (2+3+5+8+34+89+144),

(1+2+3+5+8+377),

(1+2+3+5+8+144+233),(1+5+34+377),(1+2+3+34+377),(1+5+13+21+377),

(1+5+34+144+233),(1+2+3+13+21+377),(1+5+34+55+89+233),

(1+2+3+34+55+89+233),(1+2+3+13+21+55+89+233),

528 is the sum of 7 sets of Fibonacci numbers:

(2+5+144+377), (2+5+55+89+377),

(2+5+21+34+89+377),

(2+5+55+89+144+233),

(2+5+8+13+34+89+377),

(2+5+21+34+89+144+233),

(2+5+8+13+34+89+144+233)

639 is the sum of 20 sets of Fibonacci numbers:

(8+21+610), (3+5+21+610),

(8+21+233+377), (1+2+5+21+610),

(3+5+8+13+610), (3+5+21+233+377),

(8+21+89+144+377), (1+2+5+8+13+610),

(1+2+5+21+233+377),

(3+5+8+13+233+377),

(3+5+21+89+144+377),

(8+21+34+55+144+377),

(1+2+5+8+13+233+377),

(1+2+5+21+89+144+377),

(3+5+8+13+89+144+377),

(3+5+21+34+55+144+377),

(1+2+5+8+13+89+144+377),

(1+2+5+21+34+55+144+377),

(3+5+8+13+34+55+144+377),

(1+2+5+8+13+34+55+144+377)

741 is the sum of 21 sets of Fibonacci numbers:

(8+34+89+610), (3+5+34+89+610),

(8+13+21+89+610), (8+34+89+233+377),

(1+2+5+34+89+610),

(8+13+21+34+55+610),

(1+2+5+34+89+233+377),

(8+13+21+34+55+233+377),

(1+2+5+13+21+89+144+377),

(3+5+13+21+89+610),

(8+13+21+89+233+377),

(3+5+13+21+34+55+610),

(1+2+3+34+144+233),

(1+2+3+5+8+21+377),

(1+5+13+21+144+233),

(1+2+3+13+21+144+233),

(1+2+3+5+8+21+144+233),

(1+5+13+21+55+89+233),

(1+2+3+5+8+21+55+89+233)

(3+5+34+89+233+377),

(1+2+5+13+21+89+610),

(3+5+13+21+89+233+377),

(1+2+5+13+21+34+55+610),

(3+5+13+21+34+55+233+377),

(8+13+21+34+55+89+144+377),

(1+2+5+13+21+34+55+233+377),

(3+5+13+21+34+55+89+144+377),

(1+2+5+13+21+34+55+89+144+377)

852 is the sum of 14 sets of Fibonacci numbers:

(1+8+233+610), (1+3+5+233+610),

(1+8+89+144+610), (1+34+89+144+610),

(1+8+34+55+144+610),

(1+8+89+144+233+377),

(1+3+5+34+55+144+610),

(1+3+5+89+144+233+377),

(1+8+13+21+55+144+610),

(1+8+34+55+144+233+377),

(1+3+5+13+21+55+144+610),

(1+3+5+34+55+144+233+377),

(1+8+13+21+55+144+233+377),

(1+3+5+13+21+55+144+233+377)

963 is the sum of 12 sets of Fibonacci numbers:

(2+8+21+89+233+610),

(2+3+5+21+89+233+610),

(2+8+21+34+55+233+610),

(2+3+5+8+13+89+233+610),

(2+8+21+34+55+89+144+610),

(2+3+5+21+34+55+89+144+610),

(2+3+5+8+13+34+55+89+144+610),

(2+3+5+8+13+34+55+89+144+233+377),

(2+3+5+21+34+55+233+610),

(2+3+5+8+13+34+55+233+610),

(2+8+21+34+55+89+144+233+377),

(2+3+5+21+34+55+89+233+377),

When the notes of the Solfeggio scale are played together, you cannot create any dissonance, because they all contain the same harmonies within them. This may not be evident at first, but once our ears are "socialized" to the new sounding scale, the harmonies can be heard. This allows the Music Therapist/Sound Therapist to play any two notes together, or to strum all strings simultaneously. Since all nine notes are in tune with each other, there is no need for theory rules to follow. There are no sharps or flats, or key signatures. The therapist has only nine notes to improvise with complete freedom. If the therapist wanted their client(s) to play along with them, the client would be able to easily lie the instrument with the nine frequencies on their table or lap and begin to play with basically no instruction. Special needs persons or groups could easily play this instrument.

Carla Rose Kelly, one of the authors of this paper, put the nine frequencies on new patented sound healing instruments and conducted a qualitative and quantitative research project with co-author Julia MacKinley.

Research on the
Nine Solfeggio Frequencies:

Previous research on these frequencies has looked at groups with anxiety, depression and pain. The study consisted of 120 participants gathered from a community sample who completed 3 sessions. During each one-hour session, they listened to love based, positive affirmations put to improvised music, utilizing the nine solfeggio frequencies. All participants were lying down on yoga mats, in a dimly lit room, with eyes closed. The sessions were one week apart. The musical instruments in the study were provided by and are proprietary of one of the authors, Carla Rose Kelly, and are described in Patent Publication Number Canada: 2,813,225; US: 9,092,993 B2; US: 9 821,137 B2 and Japan: 6034291

The research sessions included vocals, sung along to a lap harp, chimes, flutes and tuning forks all made with the nine frequencies. The instruments were used in a random order with no predetermined amount of time for each. The music was improvised for one hour. The melodies and rhythms changed at the will of the therapist, with no predetermined formula. The use of the notes was completely chosen in the moment, again with no predetermined formula.

The participants were not asked to sing along in the study. In this sample, 31% had been diagnosed with a psychiatric condition at some point in their life and 65% reported some form of chronic pain in their lifetime. All healing music sessions had a random mix of persons with emotional and physical pain. A repeated measures ANOVA with a Greenhouse-Geisser correction determined the mean reported level of depression differed significantly across sessions ($p < 0.001$). The average depression score at baseline was 10.2 out of a possible

60, whereas after the final session the average depression score was 2.5, a score noting a lack of depression. The same trend held true for anxiety ($p < 0.05$) and pain ($p = 0.01$), decreasing mean reported pain by 4.2 points on a 10-point scale and decreasing mean anxiety 16 points on a 99-point scale.

Although the research showed improvements, the testimonials provided by the participants may be even more powerful evidence of the impact of this method. Below are a selection of testimonials from previous study sessions:

"This was profound on many levels. The pain in my right quadriceps and right thumb was released immediately. Years of physiotherapy have not had effects as amazing as these sessions."

"I had low back, nerve root damage causing sciatic pain, numbness, and spasms down left leg. The sound healing sessions lessened my episodes of sciatic pain. I felt grounded and much calmer.

"I have discontinued medication for my migraines after attending the sound healing sessions. Thank you! Thank you! Thank you!"

Common experiences ranged from tingling or rushing of cold/warm energy down a limb, to what feels like pain/tensions, or stresses leaving the body. People often reported feeling much lighter after, like the weight had been lifted off their hearts, off their minds. They felt more connected to their soul. They reported clarity in the thinking, increased sensitivity of their senses. Some participants even reported spiritual connections. Overall, the reports were very positive and tended to occur in 1-3 sessions, making this type of healing very fast acting when compared to pharmacology. They reported improvements in mind, body and spirit.

This study shows that this form of treatment is effective in alleviating the symptoms of pain, depression and anxiety. When we looked at different forms of pain, the treatment was equally effective across pain types (ie. migraine, joint, back, muscle pain etc.). Also, worth noting, that persons with emotional and physical pain were listening to the same positive affirmations for each session, in that we did not separate the groups into pain types. This shows that this type of therapy can help alleviate pain across a variety of symptoms and disorders. The fact that it is a passive therapy leaves it available for people suffering from a wide array of disorders that may impact language, spatial and motor functioning.

Entrainment

Entrainment is a common term in Music & Medicine. It can be defined as a synchronization of two or more rhythmic cycles. We rely on entrainment in our Sound Therapy/Music Therapy sessions so that imbalances in the client will vibrate into sync with the music or sounds produced for them to heal or come into balance. A simple example, is when we spend time out in nature, listening to the birds, wind, rain, etc., we become entrained with those vibrations, and they make us feel good. Nature sounds are calming, soothing, healing, and rejuvenate us, bringing us back into balance. The human body is created in the same creational pattern as everything else in nature, so it entrains our mind-body-soul back to its natural state, where it feels the best. Since the nine Solfeggio frequencies contain the same harmonies as our human bodies were created with, it is easy to understand how persons listening to these frequencies will entrain themselves back into balance.

A New Solfeggio SACRED SCALE® Harp Made in the PHI Ratio

The harp used in the study was not constructed by the luthier with the PHI ratio in mind. However, it did work, producing substantial results. Since this time, a new Solfeggio SACRED SCALE® Harp was developed with the PHI ratio in mind. Not only are the frequencies in tune with the Fibonacci Series, but the entire instrument is now too. The length of the new Solfeggio SACRED SCALE® Harp is 21 inches. The sound chamber is 13 inches wide and 13 inches long. The sound hole is 3 inches diameter. Three, thirteen and twenty-one are all Fibonacci Series numbers. This instrument has been used in recent sound therapy sessions, with clients reporting it more powerful than the one used previously in the study. The following are two testimonials of their experiences with a ten to fifteen-minute session with the harp played in the room around them, and laid on top of their chest and legs.

1. I lie down and close my eyes and Carla starts to play the harp in the room around me. From the beginning, it feels like the voice and sound of the harp transforms. More

light comes in and heals my body. I felt lots of release in my right leg. Just like letting go of control and tension with each stroke of the harp, something changes, vibrates and goes into your body and soul. It feels like the power of the music and the vibrations expand over the air. It feels like the voice, sound and vibration all connects into one. Each word she sang (ex. confidence, love) becomes reality in me. When she sang love with the harp, I felt more love. When she sang confidence, I felt more confidence. The words and sound all together create healing. Then she placed the harp on my legs and played. My right leg was "letting go", vibrating. The pain of the knee went away. There was lots of heat and release. When the words and sound of the harp amplified, it felt like angels were singing. When the harp was placed directly on my heart, and strummed, there was a powerful healing of the heart to more love and peace. My head also healed. I saw lots of white light and it expands also with the sound in echoes through the air. At the end, I could see more light. It transported me to feeling like "flying in space". It transforms me into more light and love. P.C. London, Ont.

2. Angelic and peaceful sum up what the harp felt like. As soon as the first note is played, I felt surrounded by angels. I was overcome with a peaceful sensation throughout the entire body. It felt like all was well with me and the world. I felt complete relaxation and a calming of the mind. It was like my entire mind-body-soul was in harmony. From the moment it started, I had visual images throughout, starting with seeing angels. There was an emotional healing which brought in more peace and calm to the mind, especially when the harp was placed on the heart and played. I felt different body sensations and thoughts would come to mind when different affirmations were sung. I opened my eyes when the session was done and I felt more relaxed and felt 20 years younger on my face and body. It was like any and every

stressful, traumatic experience just melted away. All worries and stress disappeared. K.D. Port Stanley, Ont.

Summary

Were the six frequencies that were hidden in the Book of Numbers in the bible for over 2000 years, part of the nine frequencies that are so powerful, they create a Divine matrix for healing and union with Divine? This seems possible with their mathematical connection to creation. Why were they hidden? Are they the frequencies that were on the magical musical instruments used by David and other "physician musicians" to heal the ills of the day? Possibly, but there is no evidence uncovered in our history to know the truth at this time. It is interesting that both the story of David healing King Saul with the harp, and the hidden frequencies were together, in the most popular book in the world, the Bible. What we do know, is the nine frequencies do have healing power as evidenced by the research presented in this paper. We also know that they are in harmony with nature, with the universe, and creation itself. We understand how they entrain the mind-body-soul back into harmony, so music made with these frequencies can be used with certainty as medicine. Although the research study discussed herein only involved three, one-hour sessions, the results consistently moved the symptoms of depression, anxiety and pain towards healing. If the sessions had kept going, would those participants eventually be completely relieved from their symptoms? Further research needs to be done to determine if the nine frequencies can heal the pain, anxiety and depression symptoms completely and how many sessions would be needed for the same.

References

Fibonacci, L., & Sigler, L. E. (2002). *Fibonacci's Liber abaci: A translation into modern English of Leonardo Pisano's Book of calculation*. New York: Springer.

Grout, D. J., Grout, D. J., & Palisca, C. V. (2010). *A History of Western Music*. New York: W.W. Norton & Company.

Horowitz, Leonard G. & Puleo, JS. (2001). *Healing Codes for the Biological Apocalypse*. Publisher: Tetrahedron Publishing Group; Sandpoint, Idaho.

Jakowiak, K. (2019, May). What is attuned vibrations about? www.attunedvibrations.com.

Music and Mental Health:

A Systematic Review of Literature on Functional Music Medicine with Reference to Kenya

David O. Akombo

Abstract

This study provides an understanding of the role of music in mitigating the stigma of mental health and address the high costs of mental health in Kenya. The study focuses on psychiatric disorders because mental illness in Kenya is both of a national and international concern. The global burden of disease related to mental disorders is on the increase, with the World Health Organization (WHO) estimating that worldwide, over 450 million people are affected. In Kenya, the global mental health concern as with other parts of the world, was triggered by the Movement for Global Mental Health (MGMH) and the disagreements regarding the 5th edition of the Diagnostic and Statistical Manual of Mental Disorders (DSM-5) by the American Psychiatric Association (APA). The National Institute of Mental Health (NIMH) and the British Psychological Society (BPS) have both questioned the validity of DSM's classification. BPS went further in questioning the science behind the biomedical approach and called for a paradigm shift. Thus, the campaign to 'scale up' psychiatric treatments in developing nations itself is standing on a shaky ground, since it is primarily driven by the biomedical approach. Reforming mental health in Kenya is not only a medical matter but an issue of justice and freedom - it's about honoring those psychiatrized and treated unfairly by the colonial system. It is about decolonizing mental health.

Introduction

Patients with mental illness show significant cognitive and emotional benefits when they sing, or listened to familiar songs. Depression and anxiety are some of the most serious and frequent mental disorders worldwide. Thus, depression is one of the most common of all chronic diseases. This paper focuses on a systematic review of published literature on the subject matter of facets of mental illness, and how music in particular is used to mitigate in the illness. The first segment of this presentation will include studies on current causes of depression in selected regions of the world using the country of Kenya as a prototype. The segment will adopt a data-driven research approach on how music is used to mitigate these mental illness. The second segment of the presentation will focus on neuro-physiological parameters that respond to syntactic musical stimuli with Cross-Domain Mappings in music. The third segment of this literature will address African drumming and how it can be used in managing symptoms of mental illness. Overall, the paper focuses on the benefits of music to patients with depression and impinges upon ancillary topics on music perception, memory, and learning and their relation to the sensory, formal, and expressive properties of music.

Background of Mental Health in Kenya

Kenya has 54 medically trained psychiatrists, 418 medically trained psychiatric nurses, 10 medical social workers and very few psychologists10 to cater to a population of about 43 million, 4% of whom are likely to suffer from a major mental disorder (Ndetei et al., 2009; World Bank, 2012). The Mathari Mental Hospital located in Nairobi, was established in the late 19th Century as an isolation center for chicken pox (McCulloch, 1995). In 1910, as the need for psychiatric care grew with the growing population of White settlers both in Kenya and neighboring colonies such as Uganda, and Tanzania, the center was renamed 'The Nairobi Lunatic Asylum' (McCulloch, 1995; Mahone & Vaughan, 2007). This century-old colonial era facility represents the state of mental health care in Kenya from historical and contemporary viewpoints. It is still colonial form in law and practice; institutionalized, overcrowded, relegated to the margins and chronically underfunded by the government (KHRC, 2011). The award winning CNN documentary on mental health in Kenya 'Locked Up and Forgotten' showed the neglect and abuse in Mathari Hospital (McKenzie, 2011). Following the completion of the documentary, the Kenya Human Right Commission (KHRC) reviewed the state of mental health in Kenya and found that psychiatric care still remains both institutionalized and centralized - with 70% of all psychiatric in-patients located at Mathari Hospital in Nairobi (Kenya Human Rights Commission, 2011).

Musical stimuli and Neuro-physiological parameters in patients with mental illness

Research has shown that patients with mental illness show significant cognitive and emotional benefits when they sing, or listened to familiar songs (Särkämö et al., 2008, 2014). Depression is one of the most serious and frequent mental disorders worldwide. Thus, depression is one of the most common chronic diseases (Leubner & Hinterberger, 2017). The global burden of disease related to mental disorders is on the increase, with the World Health Organization (WHO) estimating that worldwide, over 450 million people are affected (Marangu et al., 2014). The global mental health concern was triggered by the Movement for Global Mental Health (MGMH) and the disagreements regarding the 5th edition of the Diagnostic and Statistical Manual of Mental Disorders (DSM-5) by the American Psychiatric Association (APA). The National Institute of Mental Health (NIMH) and the British Psychological Society (BPS) have both questioned the validity of DSM's classification. BPS went further in questioning the science behind the biomedical approach and called for a paradigm shift. Thus, the campaign to 'scale up' psychiatric treatments in developing nations itself is standing on a shaky ground, since it is primarily driven by the biomedical approach (Summerfield, 2008; BPS, 2011; Insel, 2013). Reforming mental health in the world is not only a medical matter but an issue of justice and freedom - it's about honoring those psychiatrized and treated unfairly by the colonial system. It is about decolonizing mental health (Ibrahim, 2014).

A larger percentage of people across the African continent including almost 80% of Kenyans, still utilize complementary and alternative medicines (CAM) including traditional and faith-based healing systems for mental health. However, the efficacies of CAM have been put into question by the Government's formal health care system (Mbwayo et al, 2013). It is understood that various herbs, like Rauwolfia, are widely in use (and were used even before the colonial period), and different ethnic groups utilize unique healing systems that are accessible and affordable to community members, and have been used by generations after generation. Yet these systems remain understudied and unrecognized by formal mental health care (Incayawar, Wintrop & Bourchard, 2009; Namachandran, 2011; Njenga, 2002; Mbwayo et al, 2013). According to a study done by Mbwayo et al. (2013), in three densely populated settlements in Nairobi, the majority of the residents utilize traditional healing systems, and cited overall satisfaction with traditional treatments, flexible payments and easy accessibility, as some of the reasons for choosing healers over the mainstream health care services. Other studies (Gesler et al, 1995; Abbo, 2011) in the neighboring countries of Uganda and Tanzania reported similar outcomes.

Mental health costs in Kenya are influenced by market and non-market factors that determine pricing, along with possible causes as to why the ordinary Kenyans who need mental health are unable to access it. About 82% of the total land in Kenya is classified as arid and semi-arid (Haradhan, 2013). Kenya's per capita healthcare spending is USD 1,200 which is slightly higher than its gross national income (GNI) of $1,160. This means that an individual with mental health needs will have to spend all their income on mental health at the expense of their subsistence needs (World Bank, 2016).

This research paper attempt to inform the field of CAM which suggests that music making and listening can be used effectively as a modality for intervention of mental health conditions due to proven of potential music for stress reduction (Pelletier, 2004), emotional regulation and personal mood management (DeNora, 2000; Hays & Minichiello, 2005; Sloboda, 1999), pleasure and the arousal of positive emotions (Gabrielsson, 2011; Laukka, 2007; Salimpoor, Benovoy, Larcher, Dagher, & Zatorre, 2011), social bonding (Nilsson, 2008), and enhancement of cognitive functioning after neurological impairment (Särkämö, Tervaniemi, & Laitinen, 2008). This intervention is generally low in cost and is readily available to clients. Any care giver or educator, with some form of training in the use of music performance to alleviate symptoms of mental health illness could offer this CAM to patients. Clients should be encouraged to use music listening in a place where other pain and anxiety mitigating methods are likely to be minimal. In hospitals, offering music intervention as a coping resource significantly decreased pain, anxiety, and physiological variables, such as blood pressure, heart rate, and respiratory rates.

This project represents a groundbreaking initiative to bring the arts to healthcare settings in Kenya in the future. In Kenya, well over two million people are affected mental health illness, and poverty is a major determinant in the high numbers of people living with mental health illness (Marangu et al., 2014). These statistics, along with other significant health, social, and economic issues, make Africa an area of extraordinary need. In highly industrialized nations including the United States and Great Britain, over the past three decades, extensive research has shown the efficacy of the arts in enhancing physical and emotional health and in improving the experience of healthcare for patients, their families, and caregivers. Not only do these programs show measurable health benefits, but they also establish healthcare centers as cultural centers, delivering the arts to places where people of diverse cultural and socio-economic backgrounds are institutionalized. Africa is a continent where the arts still play significant roles in everyday life but are entirely absent in hospitals where more and more people are spending significant amounts of time due to personal or family illness.

Music Therapy is the clinical and evidence-based use of music interventions to accomplish individualized goals within a therapeutic relationship by a credentialed professional who has completed an approved music therapy program. (American Music Therapy Association, 2019). The approach of using music therapy to address mental health illness is consistent with the National Institute of Mental Health's Behavioral therapy which is used to treat avoidance symptoms. Behavior techniques used include the graded exposure to rhythms and melodies a like. Just like the medications act upon the central nervous system to reduce feelings of anxiety and associated ailments, so the drums elicit that effect. Antidepressant medications have been proven effective in treating PTSD, including newer compound agents such as Desyrel (trazodone), Xanax (alprazolam), or Lexapro (escitalopram). Sedating agents can help with sleep disturbance. Anti-anxiety agents may be useful, but the benzodiazepines, a common class of these medications, can potentially become as potent as music.

African Drumming and its Effect on Mental Health

The urge to move to music is universal among humans (Levitin, Grahn & London, 2018). Music and movement evoke a wide range of feelings in the human body, mind, and soul. Feelings (along with emotions, the phenomena that are commonly and unfortunately taken as their equivalent) are only considered from psychological or sociocultural perspectives, without consideration of the fact that feelings are also neurobiological phenomena and play a central role in homeostasis (Sloboda, O'Neill, & Ivaldi, 2001; Craig, 2002). When listeners get carried away by the music, either through movement (such as dancing) or through reverie (such as trance), it is usually the temporal qualities of the music—its pulse, tempo, and rhythmic patterns—that put them in this state (Levitin, Grahn & London, 2018). This experiential workshop will involve participants listening to both Crystal bowls and drum sounds while making voluntary movements, chanting, humming, and experiencing the presence of the cosmos within the human mind, body, and soul.

Research has shown that African drumming offers psychological benefits by encouraging social interaction, building skills that can make a meaningful contribution to society. These benefits can be gleaned from the works of (Longhofer and Floersch, 1993), which examine the music and the promotion of competency in participants. Other areas of research in the field include music-induced facilitation of self-expression, non-verbal communication and improved physical, emotional and mental health (Kalani, 2004); and musically induced reduction mechanism of negative emotions such as anger and frustration (Hull, 1998). Drumming is rapidly becoming recognized and utilized as a medium for personal growth and development, for the enhancement of community feeling and for physiological and psychological healing (Friedman, 2000:3). Longhofer and Floersch (1993:3-8) specifically describe the value of community drumming practices originating from Ghana, West Africa, for psychiatric rehabilitation at a mental health center (as described in greater detail in Chapter Three). Psychotherapist and drumming facilitator Robert Lawrence Friedman (2000:4) confirms that "There is a burgeoning body of scientific literature which points to the drum's ability to promote well-being." He continues (2000:6): "Yet much more research is necessary to confirm what many have experienced and studied already – that rhythm in relationship to the body and mind has many benefits – an idea that our society is just beginning to grasp." In her inspirational study in South Africa, Flores (2011), found that African drumming influenced the children's levels of self-concept, depression and disruptive behavior; similarly, age seemed to interact with both anxiety and disruptive behavior to produce observable outcomes.

Conclusion

This paper has shown that the existing literature indicated that both listening to and creating music can have various positive effects on mood and mental health. The use of music for promoting good health, for healing and for maintaining homeostasis has existed in many cultures throughout history. Even though diagnostic, prescriptive and prognostic models have changed over the centuries, the ancient view on healing as holistic, involving intimate interactions between body, mind, environment and spirit has not yet fully changed. According to this view, human beings throughout all societies view the use of music as an intelligent, conscious and structured method to mitigate in the area of mental health. Mental health seems to require holistic healing in which the process requires multiple and interlinked biological, psychological and social factors. Holistic healing therefore encompasses well-being in all aspects of the person – the physical, psychological, emotional and spiritual which are all intertwined with the cognitive domain. Music for mental health is understood in all aspects of art forms that include, but are not limited to, singing, chanting, yodeling, dancing, and playing of musical instruments for therapeutic purposes. The desire to explore how music is used for mental healthcare delivery, to alleviate suffering, and to produce less violent and more peaceful outcomes for those most disadvantaged in society motivated this study.

References

Abbo, C. (2011). Profiles and Outcome of Traditional Healing Practices for Severe Mental Illness in Two Districts of Eastern Uganda. Global Health Action, 4, 711.

American Music Therapy Association (2019). What is Music Therapy? 8455 Colesville Road, Suite 1000, Silver Spring MD 20910 https://www.musictherapy.org/about/musictherapy/

British Psychological Society. (2011). Response to the American Psychiatric Association: DSM-5 Development. [Online] The British Psychological Society, 2011, 1-26. Accessed 27 June 2019 at: http://apps.bps.org.uk/_publicationfiles/consultation-responses/DSM-5%202011%20-%20BPS%20response.pdf.

DeNora T. (2000). *Music in everyday life.* Cambridge: Cambridge University Press.

Flores, K. (2011). *African drumming as a medium to promote emotional and social well-being of children aged 7 to 12 in residential care.* Unpublished Dissertation. Pretoria, South Africa: University of Pretoria.

Friedman, R. L. 2000. *The Healing Power of the Drum*. Reno: White Cliffs Media.

Gabrielsson A. (2011). *Strong experiences with music.* Oxford: Oxford University Press.

Gesler, M., Msuya, E., Nkunya, M., Schar, A., Heinrich, M.,Tanner, M. (1995). *Traditional Healers in Tanzania: Social Cultural Profile and Three Short Stories*. Journal of Ethnopharmacology. 48, 145-160.

Hays T., & Minichiello V. (2005). *The meaning of music in the lives of older people: A qualitative study*. Psychology of Music, 33(4), 437–451.

Hull, A. (1998). *Drum Circle Spirit: Facilitating Human Potential through Rhythm*. Reno: White Cliffs Media.

Incayawar, M., Winthrob, R., Bourchard, L. (2009). *Psychiatrists and Traditional Healers: Unwitting Partners in Global Mental Health.* West Sussex: World Psychiatric Association.

Wiley-Blackwell. Insel, T. (2013). *Transforming Diagnosis*. National Institute of Mental Health. (NIMH). Accessed 26 June, 2019 from: http://www.nimh.nih.gov/about/director/2013/transforming-diagnosis.shtml.

Kalani, D. (2004). *Together in Rhythm: A Facilitator's Guide to Drum Circle Music*. Los Angeles: Alfred Publishing.

Kenya Human Rights Commission. (2011). *Silenced Minds: The Systemic Neglect of Mental Health System in Kenya*. [Online] The Kenya Human Rights Commission. Available at: http://www.knchr.org/. (Accessed 24 June 2013).

Laukka P. (2007). *Uses of music and psychological well-being among the elderly*. Journal of Happiness Studies, 8(2), 215–241.

Levitin, D. J., Jessica A. Grahn, and Justin London (2018). *The Psychology of Music: Rhythm and Movement*. Annual Review of Psychology, 2018. 69:51–75.

Mahone, S. (2006). *The Psychology of Rebellion; Colonial Medical Responses to Dissent in British East Africa.* The Journal of African History, 47(2), 241-258.

Mahone, S., Vaughan, M. (2007). *Psychiatry and Empire*. Hampshire: Palgrave Macmillan.

Marangu E, Sands N, Rolley J, Ndetei D, Mansouri F. (2014). *Mental Healthcare in Kenya: Exploring optimal conditions for capacity building.* African Journal of Primary Health Care and Family Medicine, 6(1): 682-6. Accessed July 31, 2019 from http://dx.doi.org/10.4102/phcfm.v6i1.682.

Mbwayo, A., Ndetei, D., Mutiso, V., Khasakhala, L. (2013). Traditional Healers and Provision of Mental Health Services in Cosmopolitan Informal Settlement in Nairobi, Kenya. African Journal of Psychiatry. 16, 134-140.

McCulloch, J. (1995). *Colonial Psychiatry and "the African mind"*. Cambridge: Cambridge University Press.

McKenzie, D. (2011). *Kenya's Mentally Ill Locked Up and Forgotten.* [Online] CNN World. Accessed August 3, 2019 from http://www.cnn.com/2011/WORLD/africa/02/25/kenya.forgotten.health/index.html.

Mohajan, H. K. (2013). *Poverty and Economic Development of Kenya*. In International Journal of Information Technology and Business Management Vol.18 No.1: 72-76.

Mohamed, Ibrahim (2014). *Mental Health in Kenya: Not yet Uhuru.* In Disability and the Global South, 2014 Vol.1, No. 2, 393-400.

Ndetei D, Khasakhala L, Kuria MW, et al. *The prevalence of mental disorders in adults in different level general medical facilities in Kenya: A cross-sectional study.* Ann Gen Psychiatry. 2009; 8:1. http://dx.doi.org/10.1186/1744-859X-8-1

Njenga, F. (2002). *Focus on Psychiatry in East Africa.* British Journal of Psychiatry, 181, 354-359.

Nilsson U. (2008). *Effects of music interventions: A systematic review*. AORN Journal, 87(4), 780–807.

Salimpoor V. N., Benovoy M., Larcher K., Dagher A., & Zatorre R. J. (2011). *Anatomically distinct dopamine release during anticipation and experience of peak emotion to music.* Nature Neuroscience, 14(2), 257–355.

Särkämö T., Tervaniemi M., & Laitinen S. Y. M. (2008). *Music listening enhances cognitive recovery and mood after middle cerebral artery stroke*. Brain, 131, 866–876.

Sloboda J. A. (1999). *Everyday uses of music listening: A preliminary study.* In Yi S. W. (Ed.), Music, mind, and science (pp. 354–369). Seoul: Western Music Institute.

Sloboda, J. A., O'Neill, S. A., & Ivaldi, A. (2001). *Functions of music in everyday life: An exploratory study using the experience sampling method*. Musicae Scientiae, 5, 9 –32. doi:10.1177/102986490100500102

Steinbeis, N., Koelsch, S., & Sloboda, J. (2006). *The role of harmonic expectancy violations in musical emotions: Evidence from subjective, physiological, and neural responses.* Journal of Cognitive Neuroscience, 18, 1380 –1393. doi:10.1162/jocn.2006.18.8.1380.

Summerfield, D. (2008). *How Scientifically Valid is the Knowledge Base of Global Mental Health?* British Medical Journal, 336(7661), 992-994.

World Bank. *Kenya at a glance* [document on the Internet]. c2012 [cited 2014 Aug 31]. Available from: http://devdata.worldbank.org/AAG/ken_aag.pdf

World Bank (2016). *Economic Memorandum: Kenya's Growth Story: Past, Present and Future.* Wahsington D.C.: The World Bank Group. Accessed July 26, 2019 from https://www.worldbank.org/en/country/kenya/publication/kenya-economic-memorandum-from-economic-growth-jobs-shared-prosperity

World Health Organization (1946). *Constitution of the World Health Organization.* Geneva: Author.

World Health Organization. (2008). *Scaling Up Care for Mental, Neurological and Substance Use Disorders.* Mental Health Gap Action Program. Geneva: WHO.

The Effect of Instrumental Performance on Diabetic Musicians vs. Non-Diabetic Musicians

Derrick Alan Crow

Abstract

The study explored how blood glucose and anxiety affects a diabetic musician's health, and examines the relationship of each variable, and how each interacts within the realm of musical performance. As the number of people diagnosed with Type 1 and 2 diabetes, it becomes increasingly important to understand how this directly relates to instrumental performers at all levels. The researchers' hypothesis was that instrumental rehearsal has affected the blood glucose and cortisol on a diabetic musician. This relationship was explored through the examination of existing literature on anxiety and blood glucose as they relate to music, and through the mixed methods examination of 20 subjects. The blood glucose data was collected in a quasi-experimental method, and anxiety levels were gathered through likert scale survey. All data was examined for correlation between variables, between subject classifications, and against historic predictive literature.

The researcher also found significant correlations between perceived anxiety and glucose movement, and found a significant correlation between perceived anxiety and perceived performance success. The data suggested a difference between diabetic, pre-diabetic, and non-diabetic subjects. The analysis of these and relationship between variables and between participants suggest that diabetics and non-diabetics respond differently to instrumental playing at a biological level.

Introduction

The study explored how blood glucose and anxiety affects a diabetic musician's health, and examines the relationship of each variable, and how each interacts within the realm of musical performance. As the number of people diagnosed with Type 1 and 2 diabetes grows, it becomes increasingly important to understand how this directly relates to instrumental performers at all levels. The researcher's hypothesis was that instrumental rehearsal has affected the blood glucose and cortisol on a diabetic musician. This relationship was explored through the examination of existing literature on anxiety and blood glucose as they relate to music, and through the mixed methods examination of 20 subjects. The blood glucose data was collected in a quasi-experimental method, and anxiety levels were gathered through likert scale survey. All data was examined for correlation between variables, between subject classifications, and against historic predictive literature.

The act of playing an instrument performance effects the performer. We have seen spikes in Cortisol (LeBlanc, Jin, Obert, and Stivol, 1997 and Kokotaski and Davidson, 2003), muscle and hand injury (Ostwald, Avery, and Deschamps, 1991) and Blood Glucose (Crow, 2005 and 2006), but the difference between diabetic and non-diabetic musicians is a relatively new line of exploration. With the increase in childhood diabetes, hypoglycemia, and overall obesity in the American population, we need to understand how instrumental performance, and rehearsal effect the individual, and more over how playing a musical instrument interacts with a diabetic. The scope of this article will be limited to the effect of instrumental rehearsal on diabetic and non-diabetic musicians and look for general and specific correlations within and between these groups.

On an anecdotal level many performers have reported being hungry or feeling tired after performances. Some have even reported symptoms that are similar to those of low blood sugar. While these personal accounts do not prove a change in blood glucose, previous studies (Wan and Gail 2005, Kemp 1996, and Wilson 1997) have shown that musicians are aware of how their body is responding to the performance environment.

Literature Review

There are limited studies on the effects of instrumental rehearsal on blood glucose. At the time of this paper, there are only two studies on this topic, and both are by this author. The first pilot study in 2005, which was presented in full at a 2005 graduate researcher symposium at the University of North Texas, and a master's thesis in 2006 and presented at the 2012 Interdisciplinary Society for Quantitative Research in Music and Medicine. We will use these two studies to define and frame the variables and concepts to be used in the comparison of diabetic versus non-diabetic musicians.

In. 2004 the original study was began due to an accident which prevented a diabetic musician from their normal practice routine and inspired the initial question. What if any effect does instrumental rehearsal have on blood glucose levels? A heterogeneous sample of 20 musicians were tested four times. Once for the control and three times for the treatment variable, and all draws were done in similar conditions to limit confounding variables. The study found correlations of P .05 or higher in the relationship between perceived anxiety and perceived performance success, and a significant relationship was found between perceived anxiety and glucose level movement. The examination of data found new confounding variables and presented interesting trends in glucose movement that required further study. This led to a replication of the study, which delimited some of the confounding variables and focused more on the effort over time of the musician.

Crow (2006) was a case study of five low brass musicians over several months. This study looked at performers with similar static load, performance requirements, and breath control. The study showed a significant correlation between perceived anxiety and glucose movement, it also shows a significant relationship between perceived anxiety and perceived performance success. The study did offer a clearer picture of the effect of instrumental rehearsal on the performer and presented interesting outliers that offer new areas of research. There was a noticeable difference between the diabetic, pre-diabetic, and the non-diabetic participants. This suggested the act of playing an instrument effects these groups in differing ways.

Methodology

Subjects were drawn from a performance ensemble at a Midwestern university. The subjects were tested with a blood glucose meter twice a day once a week over the span of five weeks, and the testing occurred once before rehearsal and again after rehearsal had ended. Glucose levels were also checked on a non-rehearsal day to establish a base line for comparison.

The subjects were recruited through a classroom presentation of the planned experiment. The potential participants were asked to volunteer after the methodology and requirements were explained. The volunteers signed an interest form stating a willingness to be part of the study. Each volunteer was assigned a number. From these volunteers twelve subjects were picked randomly. The twenty subjects were assigned a new number between 20. The researcher also participated in the study bringing the beginning N to thirteen. Through the experiment, one subject left the study due to personal reasons. The final N for the experiment was 20.

The testing took place within the rehearsal setting, and the subjects were tested in the same room they rehearse in. The baseline day was the only exception. The researcher tested the subjects in a setting of subjects' convenience.

The subjects had their glucose checked prior to rehearsal. The subjects would first have the area for testing cleaned with an alcohol swab. A new lancet was loaded into the penlet. The subjects were then pricked with the penlet. The Reli/On glucose meter was loaded with a new test strip. A drop of blood, 2.5 ul, was placed on the test strip. The area from which the blood had been drawn was cleaned with another alcohol swab. The glucose meter reading was recorded. The subjects preceded with normal rehearsal activities. After rehearsal, the subjects' glucose was again checked using the same process as previously described. The results were recorded and then compared to the earlier readings. The difference of the two readings was documented. The subjects were also asked to respond on a likert scale about the perceived stress and performance achievement of that day's rehearsal. The scale was 1—5, one being the least performance achievement or perceived stress and five being the maximum amount of stress or achievement.

The baseline for comparison was drawn in a similar manner. The subjects' blood glucose was measured on a non-rehearsal day. The subjects had their glucose checked at relatively the same time as start of rehearsal, and it was checked again after an hour and a half. During this span of time, the subject did no physically or mentally strenuous activity. Both readings were recorded and compared to establish the baseline movement for each subject.

The resulting movement in each of the subject's glucose levels was compare to their corresponding baseline movement. The results were graphed to give a visual representation of the movement on each test day as it relates to the normal movement on a non-rehearsal day. A standard deviation was found for each of the participants. The glucose movement and likert scale responses were compared for correlation through use of a chi-square.

Results

The descriptive data for all subjects is as follows: The range of the glucose movement for rehearsal days was -110 to 44, with a mean of -11.09 and a mode of -11. The standard deviation for glucose movement on rehearsal days was 23.70. The glucose movement for non-rehearsal days was also plotted on each graph as a baseline. The range of baseline glucose change was -48 to 16 with a mean of -7.57 and a standard deviation of 18.92.

For diabetic musicians the results showed a range of glucose movement on rehearsal days for the diabetic musicians were -55 to 24 with a mean of -29, and no repeated mode. The standard deviation for rehearsal days was 34.41. The range glucose changes for non-rehearsal days was -40 to 9 with a mean of -12, and a standard deviation of 20.37.

The participants' results were analyzed using Pearson Product Moment correlations, first applied to perceived stress and glucose movement. A correlation of 0.49 was found as shown in Figure 6. The subjects' glucose movement decreased as his perceived stress increased.

The non-diabetic results showed a range of glucose movement on rehearsal days for the participants were -28 to 20 with a mean of -9.27. The standard deviation for rehearsal days was 11.89. The range on non-rehearsal days was -8 to 13 with a mean of 0.75 and a standard deviation of 8.96

The results were analyzed using Pearson Product Moment correlations. A correlation of 0.51 was found between perceived stress and glucose movement. An upward trend is visible in glucose levels as perceived stress increased for these subjects.

Discussion

The null hypothesis of the study was: Does instrumental rehearsal affect blood glucose levels differently between diabetic and non-diabetic? From this study, it cannot be definitively answered. However, many of the results in this study point to that possible conclusion. A downward movement in blood glucose was observed for most subjects, but the rehearsal-day glucose movement did not show constant, traceable movement, and sometimes showed a contrary movement to projected expectations of the physical performance demands. All subjects did demonstrate drops in blood glucose levels in line with and above their baseline glucose changes.

As expected, the participants demonstrated a highly significant relationship between the variables of perceived performance anxiety and blood glucose movement, and perceived performance success and perceived performance anxiety. This is an affirmation of the two previous studies results and suggest a more generalizable result. These correlations did not differ between the two groups, and suggest anxiety is not the variable that cause a difference between groups.

While a relationship was found between instrumental rehearsal and blood glucose movement, the correlation requires more replications and a larger *N.* Even before correlational analysis a trend is an apparent and suggests a need for further experimentation. The data does suggest a difference between how a non-diabetic and diabetic are affected by instrumental rehearsal. There is a larger glucose drop in diabetics, and this movement can be as significant and a 100 millimoles per liter. A causal link cannot be established at this time due to the presence of confounding variables and limited *N.*

The question that is presented by this data is "Why"? Why is there a difference? The diabetic musician is more sensitive to the act of playing a musical instrument. You see more variance in the movement within subjects in previous studies, and large variance between diabetic subjects in this replication. The researcher cannot conclusively draw a causal link between the biological variable and the act of performance. Continued repetition and refinement is necessary to draw a clearer understanding of the causation if any of these variables.

At the termination of the pilot study, the benefits and drawbacks became clearer, revisions were made to existing protocols, and new protocols were added. At the conclusion of this study, other changes for future experiments also became clear. The changes suggested are strictly to increase the validity and reliability of future experiments and offer more correlating factors to increase the understanding of demands on musical performers.

The format for testing the subjects must be reevaluated. A rehearsal is the most natural setting for the performer, but it presents an uncontrolled environment. A controlled test may produce a truer representation of playing an instrument on blood sugar. The subject would also have to play something that required them to be physically active, but not stressed. An option would be to have the subject perform scales in several formats for 90 minutes so that every subject would be playing exactly the same thing every time they tested. This should eliminate internal and external stress during testing.

In future studies it would be beneficial to add more bio-monitoring devices. An automated sphygmomanometer would increase clarity in the experiment by observing and recording blood pressure. Increases in blood pressure can indicate an increased workload. A pulse rate monitor offers another level of experimental control. Like the increase in blood pressure, an increased pulse rate can indicate that music is a physical activity. While both of these devices would add new information to further study, they also magnify intrusiveness. A sphygmomanometer attaches to the arm and at given intervals inflates and uses pressure against the veins and arteries to determine the blood pressure. Having this device hanging from an arm is obtrusive, and the increased pressure may affect the playing. The advantages and disadvantages of these bio-monitoring devices should be carefully weighed before they are implemented.

Conclusion

As with the previous replications the research has narrowed the confounding variables but has increased the number of variables that influence and correlate with blood glucose movement. With the change in student demographics and overall health, the topic of the effect of instrumental rehearsal on blood glucose requires continued study and refinement. Currently a trend is becoming apparent, but without a larger *N* and greater delimitation of variables not directly related to the hypothesis, we will not know the full extent of these interactions. In the previous studies we have seen movement in blood glucose is related to instrumental rehearsal, and this study presents the same trends and movements. It is worth understanding purely for scientific curiosity, but it is imperative to continue this line of study to increase the safety and well-being of the students we are charged to educate.

References

Crow, D. (2005*) The Effect of Instrumental Rehearsal on Blood Glucose Levels*. Unpublished manuscript, Southern Illinois University Carbondale.

Crow, D. (2012) *The Effect of Instrumental Rehearsal on Blood Glucose Levels of Five Low Brass Players.* Unpublished Thesis, Southern Illinois University Carbondale.

LeBlanc, A., Jin, Y., Obert, M., Siivola, C. (1997*) Effect of Audience on Music Performance Anxiety*. Journal of Research in Music Education. 45.3.480-496.

Kemp, A. (1996) *The Musical Temperament: Psychology and Personality of Musicians.* New York, New York. Oxford University Pres.

Kokotsaki, D., Davidson, J. (2003) *Investigating Musical Performance Anxiety among Music College Singing Students: A Quantitative Analysis.* Music Education Research. 5.1.45-59

Ostwald, P., Avery, M., Deschamps, L. (1991) *Musculoskeletal Problems in Instrumental Musicians*. Performing Arts Medicine. New York, New York. New York Press

Wan, C., Huon, F. (2005) *Performance Degradation Under Pressure in Music: An Examination of Attentional Processes.* Psychology of Music. 33.2.155-172

Wilson, G. (1997) *Performing Anxiety* in: D.J. Hardgraves & A. C. North (Eds). *The Social Psychology of Music.* New York, New York. Oxford Press 229-243

Influence of Music on Student Psychophysiological Well-Being

Sabina Vidulin, Iva Milovan Delić & Jasna Valić

Abstract

History teaches that the art of music has always contributed significantly to human healing and there is a vast body of recent literature that supports it. In this paper the authors investigate if listening to music of various genres (pop, rock, jazz, traditional, folk, art), frequency and duration, brings benefits to students' psychophysiological health. The results show that music plays an important role in students' life: the habit of listening to music is frequent and impacts their personal, as well as social identity. The results also show that listening to music, with certain exceptions, positively effects students' physical, mental and emotional condition. The results are interpreted within the musical, medicinal, historical and social context.

Introduction

"Since the humankind has existed, shamans of old ages performed their healing rituals in wild dances followed by percussion rhythms and songs, as it can be seen today in some African and South American tribes and cultures" (Breitenfeld, et al., 2011a, 57). From Ancient Greek philosophers and healers who accentuated the importance of the medical use of music, until the scientific development of medicine in the second part of the 19th century, music was, with more or less zeal[1] – regarding the socio-economic and political circumstances – recognized as potentially beneficial to human health. After several decades of unpopularity of music as a means of healing due to the medicinal focus on localised diseases and popularity of the surgery of a particular body area, in the last few decades we have witnessed a rise in the holistic approach to health, and music as a means of recovery forming an important part of it. Moreover, in the past decades we have seen a wider interest in music therapy to address the physical, emotional, cognitive, and social needs of patients of any kind. Today it has been vastly known that music provokes a faster recovery from many debilitating conditions.

Some historians point out that it was the World War II that brought interest in the use of music as a therapy for a number of physiological and psychological conditions. A study titled *Music for the injured soldier: A contribution of American women's military bands during WWII* (Sullivan, 2007) examines on what way and to what extent music helped to assist physical and psychological reconditioning and resocialisation of the soldiers who, in military hospitals during the war required any type of help. It is argued that American Service Forces officially recognized music as an agent capable of helping the mentally and physically wounded, and music was during the war, and its' aftermath, used as a mean of physical, psychological, cognitive and emotional recovery of many soldiers in need. This is considered the beginning of a modern music therapy.

[1] Medicine and music were vigorously connected in Renaissance, when the idea of humoral physiology (of four body humors; blood, yellow bile, black bile and phlegm) explained bodily functions, and medical endeavours to restore the balance of the four humors found a parallel in the harmony of music, expressed by the English advocate of the scientific method, Francis Bacon (1561–1626), "...the poets did well to conjoin music and medicine, in Apollo, because the office of medicine is but to tune this curious harp of man's body and to reduce it to harmony" (Breitenfeld et al., 2011a, 57).

Music therapy as a method for benefitting the health is nowadays flourishing in the societies that recognized the scope of its achievements, and can afford the implementation of this therapy in governmental health care programmes. Literature about music therapy is in the process of growing, covering myriad of themes regarding benefits of music on various illnesses and health conditions, like cerebral palsy, intellectual disability, epilepsy, mental health illness, including anxiety, depression, eating disorders, substance abuse, schizophrenia, paranoia, psychosis, and personality disorders, ataxia, aphasia, pain, sickness and terminal illnesses, HIV, AIDS, attention deficit hyperactivity disorder (ADHD), autism spectrum disorders, cardiovascular diseases, stroke, traumatic brain injuries, dementia, Parkinson's disease, and other neurological conditions (McCarthy, 2016). Music therapists, certified professionals who have completed an approved music therapy programme, customize their approach to each patient as they develop therapeutic relationships (Raedler, 2014). In 1958 Vienna music academy was the first in Europe to open the courses of music therapy, and this step was followed by other European cities as London, Hamburg, Wrocław and the others.

The foundations of music as a therapy in Croatia

In the several last decades, the Croatian Association for Music Therapy and the Croatian Medical Music Association have been promoting and developing the field of music therapy (Breitenfeld et al., 2011a). The male chamber choir Physicians Musicians of Zagreb is, along with cultural activities, engaged also in professional and scientific activities with already five international symposiums organized on the topic medicine and music. Health Centre for Artists was established at the University Hospital Dubrava. University hospital "Sisters of Mercy"[2] has already organized more than 20 international symposiums in the field of music therapy.

The first documented use of music as a remedy in Croatia dates back to 17th century when physician, professor of anatomy and theoretical medicine, Gjuro Baglivi described the effect of tarantula spider poison on the human body and

[2] Klinički bolnički centar „Sestre milosrdnice".

recommended a wild dancing style as a potential antidote, in his book *De praxi medica* (1704)[3]. Author in this work argues that the fast pace melodies i.e. vibration of the instruments are transmitted to patient's consciousness and thus encourage him to move his extremities or to jump and dance, which causes that the poison through produced sweat gets removed from the body (Breitenfeld et al., 2011b). Baglivi also points out that music in the same time affects the patient's brain which had started to paralyze due to poison, by activiating the brain function, enhanting its vital activity and thus helping the patient's condition to normalize.

Not long after, composer and organist Julio Bajamonti in his work *Il medico e la musica* (1796) wrote about physicians and music. By explaining the connection between music and medicine he advises a physician to apply music in therapy, and states that, when treating a patient, art and humanity are just as important as professionalism. He was the most important Croatian physician – composer, also called »Croatian Mozart« and his works are an essential part of the Croatian music heritage (Breitenfeld et al., 2011b). As for music to be an important help in the treatment of mental illness noticed Ivan Krstitelj Bettini who, in his Ph.D. thesis *De musicae influxu in morbis usu* (1833), analyses physiological and therapeutical benefits of music on human mental and overall health, pointing out that there is no illness that can not be alleviated by using music, and many can be sucessfully cured (Breitenfeld et al., 2011b). From this Enlightement-period papers onwards, there are dozens of professional and scientific papers in Croatia that, from many angels, argue interesting relations of music and medicine.

Fifty years ago this field was further elaborated and developed by D. Breitenfeld. It is by his merit that today in Croatia most interesting fields of medicine and music are studied and scientifically documented. In 2008 dr. Breitenfeld published the first Croatian manual *Helping children by music. Music therapy*, and in 2011 a textbook of the title *Music therapy* (Breitenfeld et al., 2011a, 59). For several decades Croatian Association for Musical Therapy and the

[3] http://hbl.lzmk.hr/clanak.aspx?id=1057 (5/18/2019)

Croatian Association for Medical Music have been promoting and developing this new area. Though there were an attempts to introduce Music therapy into university programs, systematic introduction, neither on the university nor in the governmental health care, were not sucessfull, and as a valid therapy it is sparingly developing by the efforts of the diligent individuals in the private sector.

As part of the Croatian Scientific Society for the History of Health Culture, which operates at the Faculty of Medicine, University of Rijeka, in 2006 an informal group was established with the aim of bringing together students of medicine with music and performing at various music events. They have performed in Rijeka, Pula, Labin, Karlovac, Krk, Rab and Crikvenica, and in cities outside Croatia such as Ljubljana and Izola. Humanitarian concerts have also been organized (Nadarević et al., 2017).

Students' Portrait:
Psychophysiological Features

The academic environment directly dictates: a certain age of respondents, stable physical and mental health, and to a certain extent, stable home. Indirectly, students own a certain kind of maturity, ability of concentration and adaptation, and strong will, which all allows them to go through the studies. At the same time, we could anticipate their leisure and free time, in which they could or could not listen to music, to explore if music and to what extent influences their psychophysiological well-being.

Although at first glance it is pretentious to observe the student population as a social elite, given the context of structural changes in Croatia and the challenges specific to the transition countries, the analysis and interpretation of the obtained results found that social portraits and social status are key sociocultural aspects, and Croatian students are legitimately observed as an elite part of the youth population, the future “social elites" (Ilišin, 2014, 421) and bearers of the development of contemporary Croatian society. For those who are professionally involved in the youth field, this data is very important, since it gives an insight into the student population's value system. Particularly interesting are the results that point to "the high acceptance of values related to the private sphere, while the values pertaining to the public sphere are at the bottom of the hierarchy of values" (Ilišin, 2014, 427). In the value system, however, the respondents remain faithful to the traditional model, the highly valued material position, professional success and self-reliance, and the interest in culture, art and science diminishes, with "a certain decline in life enthusiasm and student interest in the world around them" (ibid.). Recent studies (Ilišin, 2008) results show that students in Croatia are mainly focused on safe career development and existence assurance, although most respondents do not have a vision of their own employability. Thus, at the ranking level of goals, students decide on three key goals at a very high percentage: employment (97.5%), graduation (97.1%) and ensuring one's existence (96.9%).

There are some other features associated with students, which additionally mark them as a specific youth, but generally social subgroup. "The first specificity is more than average social origin, i.e. the fact that students were above average born and raised in urban environments and families where parents are more educated and with higher material standards. Another specific feature is that students are preparing for high-skilled professions and that by completing their education, they will belong to the most educated part of the population, from which, in most cases, social elites are recruited for various areas of social activity. The third is the fact that the students are expected to have an even greater level of creativity and innovation than the other young people, with the already mentioned youthful vitality and flexibility, because of their background and academic achievement. Due to the above mentioned characteristics, students within the youth population can be seen as the most sensitive seizmograph of social movements and as bearers of new trends, which will expand, to a greater or lesser extent, in other segments of the youth" (Ilišin, 2008, 222).

However, it has to be noted that while student life can bring joy of learning and discovering new knowledge, excitement in welcoming new experiences, at the same time can carry the burden of stress, anxiety and even depression. Student life usually begins in late adolescence, between 18 and 22 years of age. This is a period of identity consolidation. Young people become more independent in their life, their emotional stability and self-confidence increase. At the same time students are exposed to specific challenges. This is a period of vulnerability; the biggest prevalence of psychic disorders that occur for the first time is before age of 24. Students are exposed to many additional stresses: academic demands fulfilment, financial problems, separation from primary family and the familiar social environment, and starting more stable partnership relations (Eisenberg, 2010; according to: Kurtović, 2013).

Students experience more anxiety, depression, fear of failure and problems in social relationships, as well as more psychological problems than the general population of the same age (Stallman, 2010; according to: Kurtović, 2013).

Long exposure to stress can lead to anxiety or depression, which are among the top five mental health challenges which students face, along with eating disorders, addiction and suicide (www. bestcolleges.com, 05/10/2019).

Anxiety is emotion felt by a person who feels that his/her life situation goes beyond his/her capabilities. This emotion is named after a specific feeling of "tightness" in the chest, chest burden and difficult breathing of the person involved (lat. angere = to tighten, to strangle). The name indicates that the feeling of anxiety is located in the chest (Milivojević, 2010). This is an unpleasant feeling of fear, tension and insecurity, followed by the vegetative nervous system activation. The majority of students experience anxiety only in certain situations (exams, important occasions). Just as with other emotions, anxiety becomes a problem when it disturbs a person's everyday life. A moderate level of anxiety can positively influence a person's motivation (e.g. taking an important exam), while very intensive anxiety, as well as its absence, can diminish the result (Petz, 2005).

Depression is an emotion that occurs when a person estimates that for him/her it is impossible to live in the world as it is, and there is nothing to be done about it. Depression is subjective awareness about the lasting impossibility to adjust to the world. A person usually believes that inadaptability is his/her fault, and that he/she is inadequate and worthless as human being (Milivojević, 2010). Depression symptoms can vary from mild to severe, and can include: feeling of sadness, loss of interest or pleasure in activities once enjoyed, changes in appetite and sleeping habits, loss of energy, fatigue, feeling of worthlessness or guilt, difficulty to think, concentrate or make decisions, thoughts of death. Depression can strike at any time, but on average, it first appears during the late teens to mid-20s. Women are more likely than man to experience depression (American Psychiatric Association, 2013).

Considering that this mix of emotions experienced by students through their studies can affect their overall health, they make an interesting group to be examined whether music in their leisure and free time can make an impact on their well-being.

Empirical Research

Aim of the research

The aim of the research was to determine whether listening to music during free time impacts students' physical and emotional health.

Research tasks

The tasks were to determine: 1) how often, for which purpose and what type of music students listen to during their free time; 2) students' impact on listening to music during rest/relaxation period; 3) which feelings and psychophysiological health conditions students recognize as influenced by music they prefer.

Method

Participants

The research took place at two Croatian universities: Juraj Dobrila University of Pula and University of Rijeka. The total number of participants was 90; 40 students were from the Juraj Dobrila University of Pula studying humanities and music. Regarding the University of Rijeka, there were 50 students of medicine participating in this survey. The participants were all students of the 2nd year of the baccalaureate study programme. There were 37 male and 53 female participants. Their average age was 22.3 years (SD = 5.16).

Other variables, as the completed previous education and the way of their free time spending, show the following: some of them, prior to the university study, completed secondary school[4], and parallel to it the music programme at primary or

[4] Primary compulsory school in Croatia lasts eight years (6/7 years-old – 13/14 years-old), while secondary school lasts four years (from 15 to 18 years-old).

secondary music school[5], or they attended extracurricular activities in or outside their schools. There were 23 participants who completed primary music school, 7 of them completed secondary music school and 27 attended some extracurricular musical activities[6] or music courses during their previous education. Even though half of the students study medicine, only 10 participants completed secondary medical school. As many as 63 participants respond that they spend their free time actively in various activities. In the first place they mentioned sports and recreation (26), then music (17), art and literature (8), socialising with friends, walking, hobby, work (6) as well as science and technology activities (6). No differences were found among students regarding gender, university programme, and musical education.

Variables

The basic demographic variables (age, gender, grade, universities) were collected by open type questions. The questionnaire was prepared by the authors of the article for the purpose of this research. It has four parts. The first part contains information regarding the basic demographic variables, previous education and one open question concerning activities they spend their free time on. The second part is oriented on listening to music in general with three questions and one open-ended question about how music impacts students. The third part is about listening to music during the period of relaxation and contains 6 multiple choice questions, while the fourth part explores the listening to music in the context of potential health benefits. This part comprises 17 statements in the five degrees Likert scale and 4 multiple choice questions.

Procedure

This research took place in the week from May 6th to May 10th at two universities and was conducted by 3 teachers, two at the Pula's university and 1 in Rijeka's. To the participants were explained the purpose of

[5] The basic music education is performed according to the primary school curriculum in the duration of six years. Secondary music education is performed according to the curriculum of music in the duration of four years.

[6] Extracurricular activities take place within the school premises but are separated from regular classes. Out-of-school activities are those organized by associations, clubs and similar institutions and are not under the school jurisdiction.

the research at the beginning of the class, mentioned also that the research is anonymous and voluntary, that they can give up at any time and that the data obtained will be used solely for research purposes. The students have fulfilled the questionnaire in 15 minutes.

Results

Statistical analyses were mostly descriptive (frequencies) and for comparisons the t-test or chi-square test were performed.

Ad 1) Listening to music during students' free time

The results of the question *Do you listen to music in your free time?* indicate that most students listen to music in their free time always (39) and often (39), while 8:4 students declare that they listen to music sometimes and rarely. There are no students who do not listen to music at all. No difference regarding gender, university, musical education and the way they spend their free time (actively or passively) was found.

The results regarding the question *What type of music do you listen to during your free time?* in which we proposed 21 genres[7] show us that Pop is the most frequent type of music they listen to (50), followed by Rock (35), Folk (26), Dance/Electronica (18), Classical (16). None of the participants listened to World Music and New Age. No differences were found in the top three chosen music types: 1. Pop, 2. Rock, 3. Folk regarding gender, musical education and the way students spend their free time. Although both university groups ranked Pop at the top, different music types were on the 2nd and the 3rd place: students of humanities and music ranked Rock as second and Dance/Electronica as third, while medical students ranked Folk as second and Rock as third.

The results of the question *When do you mostly listen to music?* show that students mostly do it while pausing from learning (38), while learning (21), while doing chores (16), while working (3), while training (3). Other added possibilities (while walking, traveling, before sleeping, at

[7] Pop: 50; Rock: 35; Folk: 26; Dance/Electronica: 18; Classical (artistic): 16; Rap/Hip-hop: 12; Jazz: 10; Alter/Indie: 10; Other: 10; Punk: 9; Country: 7; Soul/R&B: 7; Blues: 7; Soundtracks/Theme song: 7; Funky. 6; Religious: 6; Heavy metal: 6; Reggae: 3; Gospel: 2; World music: 0; New Age: 0.

parties, always, rarely) were mentioned by five students and while eating by four of them. No difference regarding gender, university, musical education and how students spend their free time was found.

Regarding the open question *What kind of influence does music have on you?* the answers can be put into four categories. The students highlight that music has a positive (inspiring, motivating, excellent, entertaining) influence on them (35), relaxing (25), calming (17) and that it can depend on their mood, music genres and so on (10). No difference regarding gender, university, musical education and the way students spend their free time (actively or passively) was found.

Ad 2) Listening to music during the rest/relaxation period

The results of the question *Do you listen to music while having rest and/or relaxing?* show that they often listen to music (42), always (29), sometimes (13), rarely (4) and not at all (2). No difference regarding gender, university, musical education and the way students spend their free time was found.

Table 1: Music types and number of students who listen to them

Pop: 43 **(1st)**	Dance/ Electronica: 11	Religious: 6
Rock: 27 **(2nd)**	Rap/hip-hop: 10	Country: 5
Classical (artistic): 23 **(3rd)**	Other: 10	Gospel: 4
Folk: 21 **(4th)**	Alter/Indie: 8	Punk: 4
Jazz: 18	Soul/R&B: 8	Heavy Metal: 3
World music: 15	Soundtracks/ theme song: 7	Funky: 3
Blues: 13	Reggae: 7	New Age: 0

When asked for the second time *What type of music do you listen to*, but *while having a rest and/or relaxing*, the results are as follows.

Both female and male students rank Pop as first (43), Rock as second (27) and Classical (artistic) as the third favourite music type for rest or relaxation. Female students rank Classical music as 2nd and male students as 3rd.

Regardless of their previous musical education, students rank Pop as first, Rock as second and Classical music as their third favourite music type for rest or relaxation. Students with previous music education add Jazz and Folk on the 3rd place.

Regardless of how they spend their free time, students most frequently listen to Pop music, those that spend their free time more actively put Classical music on the second place and on the third there are both Dance/Electronica and Folk, while students that spend their free time more passively put Dance/Electronica on the second and Classical music on the third place.

Students of humanities and music rank Jazz and Classical music as first, Pop and Rock music as second, and as Other music as third. Medical students put Pop music first, Folk second and Rock third on the music for rest or relaxation rank.

When asked *What characteristics do you choose music for rest and/or relaxation by?* in the first position they mentioned mood/character (50), followed by: current personal condition (state) (48), music (46), text (33), story or situation presented in/with music (28), performers (26), personal preferences (17), other (3), composer (2), personalities (1).

Male students rank mood as first, current personal condition (state) as second and music as third, while female students rank current personal condition as first, music and mood as second, and text as third most frequent elements of music relevant for rest or relaxation. Students of humanities and music rank mood as first, music as second and current personal condition (state) as third most frequent music element. Medical students rank current personal condition (state) first, music and mood second and text third. Students with previous musical education rank music first, current personal condition second and mood third, while students without previous musical education rank mood first, current personal condition (state) second and music third. Students that spend their free time more actively rank current personal condition first, mood second and music third, while students that spend their free time more passively rank mood first, music second and text third.

Regarding the question *What music features are crucial for your choice of music for rest and relaxation?* the results show that the first music characteristics which is crucial for their choice is the mood of music (55) and music as a whole/music in general (53), followed by: melody expressiveness (42) and instruments or voice components (42), rhythm expressiveness (33), and then changes in tempo (10), changes in dynamics (9), unexpected harmony flow (6), specificity of the musical piece (2). Male students rank mood first, music as a whole second and instruments or voices third, while female students rank music as a whole as first, mood as second, and melody expressiveness as third most crucial music features relevant for rest or relaxation. Students of humanities and music rank music as a whole as first, mood as second, and melody expressiveness as third most crucial music feature. Medical students rank mood first, music as a whole second and instruments or voices third. Students with previous musical education rank music as a whole and melody expressiveness first, mood second and rhythm expressiveness third, while students without previous musical education rank mood first, music as a whole second and instruments or voices third. Students that spend their free time more actively rank mood first, music as a whole second and melody expressiveness third, while students that spend their free time more passively rank music as a whole first, melody expressiveness, character, instruments and voices second and rhythm expressiveness third.

Regarding the question *How many times a week do you listen to music for rest and relaxation?* the students highlight that they listen to music five times a week and more (60), 3-4 times a week (17), twice a week (8) and once a week (5). No difference regarding gender, university, musical education and the way students spend their free time was found.

The results for the question *How much time a day do you spend listening to music for rest?* indicate that student spend one to two hours a day listening to music for rest (43), up to an hour (29), 3 and more hours (16), while 2 of them do not listen to music for rest. No difference regarding gender, university, musical education and the way students spend their free time was found.

Ad 3) Influence of music to students' feelings and psychophysiological health conditions

Regarding the 17 statements, Table 2 shows the statements students agreed more with (from 3.50 to 4.50). The statements around 2.50 and 3.49 show that students neither agree nor disagree with the same statements, while students do not agree with the last two sentences.

Statement	M	SD
I'm relaxed while I listen to music.	4.43	0.81
I feel great while I listen to music.	4.43	0.86
I completely abandon myself to music and enjoy.	4.24	0.93
Music provokes various positive associations in me.	4.09	0.93
Music helps me improve my psychophysiological condition.	4.03	1.01
Music helps me establish my inner peace.	4.02	1.09
Music helps me reduce my worries.	3.99	1.07
I have emotional reactions to music.	3.99	1.09
Music encourages my personal development.	3.96	1.20
While listening to music I don't feel any physical pain.	3.49	1.57
While listening to music I don't feel any mental pain.	3.34	1.37
Music helps me in better focusing on work.	3.24	1.24
I feel as if I were out of space and time while I listen to music.	3.39	1.37
I have physical reactions to music.	3.03	1.37
I don't think about anything while I listen to music.	2.76	1.26
If I have any minor physical symptoms (sneezing, coughing, headache), they disappear while I listen to music.	2.10	1.13
Music provokes various negative associations in me.	1.69	0.91

Table 2: Influences of Music

No differences were found regarding gender and regarding previous musical education. Medical students expressed to have more physical reactions to music ($M = 3.30$, $SD = 1.36$) than students of humanities and music ($M = 2.70$; $SD = 1.32$) ($t = 2.11$, $df = 88$, $p < 0.05$). Students that spend their free time actively express to have more emotional reactions to music ($M = 4.12$, $SD = 1.00$) than those that spend their free time passively ($M = 3.58$; $SD = 1.21$) ($t = 2.34$, $df = 87$, $p < 0.05$).

Regarding the last three questions from the questionnaire, the results are as follows. *By increased listening to music, students mostly do not notice anything*, pain remains the same (33), but besides that, 31 participants state that they forget about pain and illness with increased music listening, that music listening reduces their headache/physical pain (18), reduces other chronical pain of any source (7), headache or physical pain disappear (5), and something else (I do not feel any pain; spiritual peace, I feel better and full of positive energy; poorer hearing) which was mentioned by five students. No differences were found regarding gender, university, musical education and the way students spend their free time.

By increased music listening, students have mostly noticed the reduction of worries (53), more positive comprehension of the surroundings (45), stronger determination and focusing (41), positive attitude towards student life challenges (40), reduction of the feeling of helplessness (25). Nothing special was mentioned by nine participants and two of them mentioned something else (listening to music with lyrics/vocals worsens concentration). No differences were found regarding gender, musical education and the way students spend their free time. Significantly more students of medicine (54 %) have positive attitudes towards student life challenges with increased music listening than students of humanities and music (33 %) ($\chi^2 = 4.16$, $df = 1$, $p < 0.05$)

By increased music listening, students have mostly noticed the reduction of frustration and nervousness (65), enhancement of tolerance towards minor daily troubles (36), easier expression of positive feelings towards their surrounding (30), enhancement of love towards the people they are close to (24), improvement of their relationships with friends, colleagues or family (15). There are 14 of them who wrote that they have noticed nothing special. No differences were found regarding gender, university, musical education and the way students spend their free time.

Discussion

Ad 1) Listening to music during students' free time

As many as 78 students (out of 90 in total) declare that they listen to music during their free time always or often. This shows that music is an important part of their life. Music is all around us: on cell phones, television, radio and it is consumed in different circumstances and for different purposes. Even when it is done mostly passively, the habit of listening to music is frequent and impacts the personal, as well as social identity. Vidulin (2018, 147) highlights: "Musical experience and expression of music are an integral part of human life. They depend on the person, its correlation with music, as well as on music work which, by its structure, becomes considerably closer to or distant from the persons' taste and preferences."

Free time contexts (leisure time) and activities are extremely broad and include sport, physical activity, tourism, media, the arts, countryside recreation, and new technologies, amongst others (Naulinger, 1988). The choice of leisure activities has an influence on the psychosomatic condition of an individual, and consequently on the quality of human work. The same model is analogously applicable to the quality of academic progression and performance of university students. Authors of the study that depicts Croatian students' leisure time claim that the choice of leisure activities influences their overall development, that is, the development of their abilities, capacities, knowledge and personality (Bouillet et al., 2008). Authors agree that leisure experience plays a critical role for youth by allowing them to exercise choice and become involved in unstructured activities. They may serve as relaxation, stimulation, or play a role in self-development. Authors further explain that free and leisure time activities are freely chosen and they represent a person's preferences and that although free time is a vast concept encompassing all the possible activities outside the obligations of work in everyday communication, it is usually used to denote those activities that persons choose for their pleasure and to satisfy their personal needs for fun, culture and rest (Bouillet et al., 2008, Flammer and Schaffner, 2003; Previšić, 2000; Ruskin and Spector, 2002).

Music as one of the most chosen activity during free time, as Terry and Karageorghis (2006) stated, increases attention, generates emotions, changes, regulates and improves the mood, evokes memories, increases the working output, diminishes inhibition and stimulates rhythmic movements.

It has to be taken into consideration that nowadays, the media strongly influence the leisure preferences of young consumers, and pop music is vastly heard through radio, television and other media. That Pop is the most frequent type of music the students listen to in their free time is not surprising. Pop is the first in terms of its frequency of listening and media promotion. Blažević, Vidulin and Trajskovski (2015) highlighted as follows: „Popular music is the music 'accepted by most inhabitants for purposes of entertainment, which is widespread through the mass media, generally reflecting the taste of the masses'" (Glavan et al., 1978, 12). Tagg (2001) states that popular music was created for mass distribution and directed toward different socially and culturally heterogeneous groups of listeners."

Likewise, Rock is still in a high social position and young people like to listen to it. Pop and Rock also have a rich tradition, social legacy, and ideological and political origin. The so-called *folk music*[8] has also been present in the last two decades in Croatia, so the results were not so surprising. This is a fusion genre blended of Balkan folk music with genres such as pop, disco, rock, electronic, hip-hop and others. This genre emerged as a subculture in the Yugoslavian countryside in the 1970s before migrating to the cities in 1980s and reached its popuarity peak in the 1990s, before and during the war in the former Yugoslavia. It is not quite explicable why this genre, which is by some critics considered kitsch, is so popular among our respondents; one of the reasons could be that the text is close to the needs of mass,

[8] More about folk music see: Hofman, A. (2010). "Kafana singers: Popular music, gender and subjectivity in the cultural space of socialist Yugoslavia", *Narodna umjet*nost 47/1, pp. 141-161.

recognizable and short melodies and the pronounced dance rhythm affect the perception of this music and lead to a stronger emotional experience because people can easily identify with this kind of music.

Students mostly listen to music while doing another job or activity, usually passively, without its deeper cognitive perception, mostly for leisure, relaxation. Students listen to music when they are reluctant to learn, which means they relax by listening to music. It is also interesting that they listen to music while learning. That is how they learn with ease. Namely, although music can be a distraction, it helps some people to learn, and even improves concentration. The influence of music is certainly positive for the research participants because it affects them by motivating, relaxing, calming them.

Ad 2) Listening to music during the rest/relaxation period

As many as 71 students listen to music while having a rest always and often, which means that music is one of the most important sources of relaxation during their rest period.

When comparing how Croatian youth used their leisure time in 1986 and 1999, Ilišin (2002, 281) finds that the young in 1986 participated more in the following activities: going to the cinema and on trips and walks, listening to folk music, visiting art exhibitions, sporting events, public gatherings and theatrical shows or participating in a lottery and competitions. In socialist Yugoslavia, the leisure genre which was popular, as it seen in this example, was only folk music, so time has undergone a great change in these 33 years.

Pop is obviously good even when students take a rest, and so is Rock. Considering the two similar questions from the questionnaire regarding what type of music students' listen to, there are some differences between listening during their free time, especially during the rest/relaxation period. It can be seen that in both cases these types are included: Pop, Rock, Folk, Dance/Electronica, Classics and Jazz. In the first case, when asked what music they listened to during their free time, they stated as follows: Pop, Rock, Folk, Dance, Classics, Rap, Jazz, while during having a rest they mentioned: Pop, Rock, Classics, Folk, Jazz, World and Dance. So,

Classical Music as well the World Music are, according to results, more applicable when having a rest. World music does not appear in the first choice, nobody selected it, while during the rest, 15 students mentioned it. The novelty is that classical music, that was previously in the fifth place, appears to be the third best during the rest period. Students have obviously recognized it as suitable for relaxation. We supposed that they took into consideration the art work with not so many oscillations in regarding to music elements on which the art work is built on. Blažević, Vidulin and Trajkovski (2015, 18) state that many musical pieces "belonging to art music are familiar to the wide public, regardless of their music education, and are part of the general knowledge. (...) Some characteristics for which the auditorium remembers, recognizes and appreciates art music works are: expressed and easy to remember melodies, motoric patterns, characteristic rhythm, performers' expressed dynamic toning, a specific way of playing, soloist/chamber/orchestral sound, the character and mood of the work, virtuoso parts, interesting libretto." Jazz was on the seventh place, and now it is on the fifth, so students think that is more suitable for listening during the rest period. Jazz music was long considered dance music, but since the middle 20th century it was meant for listening more. "The basic characteristic of jazz is improvisation, but jazz is not only improvisation; it consists of parts which are set and parts which are improvised. The three most characteristic rhythmic features of jazz are syncopated rhythm, freedom of performance and a constant rhythmic pulse. " (ibid.)

According to obtained results, the work students choose for relaxation is mostly chosen, according to its mood/character, then with regard to their current personal condition/state and music in general. So the mood that music can offer as well as music itself is crucial in selecting a piece of work, but the current personal condition (state) of a person affects the final selection. One third of the respondents also mentioned lyrics as one of the predictors which impact the selection, and can affect the deeper spheres of a person's personality.

From other typical musical elements, besides mood/character of the work and music as a whole, the importance of the melody and voice or instruments (performers) is emphasized by the students. The fact is that melody is a very important part of musical identification and music is most often remembered from its melody. Likewise, pronounced, beautiful singing and / or playing are one of the first specifics that make music closer to the audience and rememberable. More than half of the respondents listen to music for relaxation frequently, even five times a week, and only a few do that once a week. This means that music makes an important part of their rest and that they are surround by music during the relaxation period. About half of the respondents listen to music for relaxation for one to two hours a day, while a smaller number listens to it more hours or does not listen at all. This can be related to daily commitments when they are not able to listen to music or they do, but not often.

Ad 3) Influence of music to students' feelings and psychophysiological health conditions

A greater value is achieved for statements which reflect the state and feelings while listening to music (I'm relaxed while I listen to music; I feel great while I listen to music; I completely abandon myself to music and enjoy.) The first three statements are close to the experience that students have regarding listening to music. It is followed by personal associations that music can provoke (Music provokes various positive associations in me.), while the other two statements are oriented to the students' own psychophysiological condition (Music helps me in improving my psychophysiological condition; Music helps me in establishing my inner peace.). It is also highlighted that a large number of students agree that music helps in reducing worries, that they have emotional reactions to music and that music encourages their personal development.

Regarding the following statements, students nor agree not disagree: While listening to music I don't feel any physical pain; While listening to music I don't feel any mental pain; Music helps me to better focus on my work; I feel as if I were out of space and time while I listen to music; I have physical reactions to music; I don't think about anything while I listen to music.

Students do not agree with the last two statements: If I have any minor physical symptoms (sneezing, coughing, headache), they disappear while I listen to music; Music provokes various negative associations in me.

Sloboda (1992), Juslin and Västfjäll (2008) stated as follows (according to: Koelsch, 2010, 131): "In most humans music can strongly affect emotion and mood, and such effects are among the main reasons to produce and listen to music". The same author states that music-evoked emotions can modulate activity in virtually all limbic and paralimbic brain structures. These structures are considered core structures of emotional processing and motivational instincts. "Music improves health and well-being through the engagement of neurochemical systems for: reward, motivation and pleasure, stress and arousal, immunity and social affiliation" (Chanda and Levitin, 2013, 179). The same authors stated that the neurotransmitters involved are respectively: dopamine and opioids; cortisol, corticotrophin-releasing hormone, adrenocorticotropic hormone (ACTH); serotonin; oxytocin.

Kemper and Danhauer (2005, 286) state as follows: "Carefully selected music can reduce stress, enhance sense of comfort and relaxation, offer distraction from pain and enhance clinical performance." Music activities involve the widely spread processing that is coupled with the release of a range of neurochemicals[9]. One of them is dopamine, the key neurotransmitter linked to reinforcement and motivation. The brain includes several distinct dopamine pathways, one of which plays a major role in the motivational component of reward-motivated behavior. The anticipation of most types of rewards increases the level of dopamine in brain.

A study found that dopamine was released at moments of highest emotional response to music measured by „chill" or shiver down the spine, as a response to a favorite piece of music. The increase of endogenous dopamine transmission was

[9] www.audiocura.com/the-impact-of-music-on-neurochemistry/ (2013)

registered by position emission tomography (PET) (Blood, Zatorre, 2001).

Emotions are usually not only accompanied by autonomic, but also by endocrine (hormonal) effects, which, in turn, affect the immune system function. With regard to music, such effects are particularly relevant when they are related to a reduction of stress or amelioration of depression and anxiety (Koelsch, 2010).

Regarding the neurobiology of stress and arousal it can be stated as follows. The biological stress response is a very complex neuroendocrine, autonomic, metabolic and immune system activity. Several neurochemicals mediate the stress response, among them glucocorticoids (e.g. cortisol) and catecholamine's (norepinephrine and epinephrine). The biological stress response is highly adaptive in the short-term, but prolonged activation of these systems has detrimental health consequences. Prolonged, elevated circulatory levels of cortisol act as neurotoxins, weakening the ability of neurons and other cells to resist injury and making them more vulnerable to the effects of toxins and the normal attrition process. All organisms seek to maintain homeostasis. Stress can be defined as a neurochemical response to the loss of homeostatic equilibrium, motivating organism to engage in activities that will restore it. Lifestyle choices that reduce stress are thought to be highly protective against diseases and music may be among these. Listening to „relaxing music" (generally considered to have slow tempo, low pitch, and no lyrics) has been shown to reduce stress and anxiety in healthy subjects (Chanda, Levitin, 2013).

Conclusion

Healing rituals with some of the elementary musical constituents (rhythm, melody; singing, playing, dancing, listening to music) were a part of the healing process since humanity has existed. Even though music healing was unpopular during certain periods because the scientific view was preferable, we can notice that the holistic approach to health has been very important for human beings in the past decades, and Croatia tries to keep up with that trend.

This research aimed to determine whether listening to music during students' free time impacts their psychophysiological health conditions. As many as 90 students from Croatia (Pula and Rijeka), attending the second year of the baccalaureate study programme at humanities, music and medicine departments were involved into the research. The results show that 78 students declared that they listen to music during their free time always or often. As many as 71 students listen to music during their free time and while having a rest. They usually listen to Pop and Rock, and mostly choose music according to the mood/character of the musical piece, then according to current personal condition/state and music in general. From other typical musical elements, besides mood/character of the work and music as a whole, the importance of the melody and voice or instruments (performers) is emphasized by the students.

Music influences students' health and well-being. The results of the research show that the greater value is achieved for statements which reflect the state and feelings while listening to music, followed by personal associations that music provokes. The results also show that students think that music helps them improve their psychophysiological conditions and that music helps reduce worries, and helps them in their personal development. The results are in accordance with Kemper and Danhauer (2005) who highlight that music promotes a sense of well-being, improve mood and decrease anxiety.

It is very important to give a chance to music to impact students' life and make them aware of the benefits that music has on their psychophysiological condition. It is necessary to involve them in various musical activities during their free time because music can help them in this very stressful period of their life as university studying is.

References

American Psychiatric Association (2013) https://www.psychiatry.org/ accessed: 5/10/2019

Blažević, I., Vidulin, S., Trajkovski, B. (2015). *The Efficiency of Exercising Pilates to different Music Genres*. Sport Science, 8(2), pp. 16-25.

Blood, A. J., Zatorre, R. J. (2001). *Intensely Pleasurable Responses to Music Correlate with Activity in Brain Regions Implicated in Reward and Emotion*. Proceedings of the National Academy of Sciences of the United States of America (2001), 98 (20), 11818-11823.

Bouillet, D., Ilišin, V., Potočnik, D. (2008), *Kontinuitet i promjene u slobodnom vremenu hrvatskih studenata (1999- 2004)* [Continuity and Changes in Croatian University Students' Leisure Time Activities (1994-2004)]. Sociologija i prostor, 46(2), pp. 123-142.

Breitenfeld, D., Bergovec, M., Lucijanić, M., Pavić, J., Miletić, J., Buljan D. (2011a). *Medicine and Music in Croatia*. Alcoholism, 47(1), pp. 57- 76.

Breintenfeld, D., Majsec Vrbanić V. (2011b). *Muzikoterapija: pomozimo si glazbom* [Music therapy: Let us help ourselves by music]. Zagreb: Music Play.

Chanda, M. L., Levitin, D. J. (2013). *The Neurochemistry of Music.* Trends in Cognitive Sciensces, 17(4), pp. 179- 192.
Flammer, A., Schaffner, B. (2003). *Adolescent Leisure across European Nations.* New Directions for Children Adolescent Development, 4, pp. 90- 103.

Glavan, D. et all. (1978). *Sondažno istraživanje publike Rock-koncerata* [Probe Research of Rock-concert Public]. Zagreb: Centar društvenih djelatnosti Saveza socijalističke omladine Hrvatske.

Guyton, A. C., Hall, J. E. (2017). *Medicinska fiziologija* [Medical Physiology]. Zagreb: Medicinska naklada.

Ilišin, V. (2014). *Sociološki portret hrvatskih studenata* [Sociological Portrait of Croatian Students]. Institut za društvena istraživanja u Zagrebu.

Ilišin, V. (2008). *Skica za sociološki portret zagrebačkih studenata: uvod u istraživanje hrvatskih studenata danas* [Sketch for the Sociological Portrait of Zagreb Student: An Introduction to the Study of Croatian Study Today]. Sociologija i prostor, 46(3/4), pp. 221-240.

Karti, P. T. (2017). *The Effect of Meditation and Music Listening on the Anxiety Level, Operation Tolerance and Pain Perception in People who were Performed Colonoscopy.* International Journal of Caring Sciences, 10(3), pp. 1587-1594.

Kemper, K. J., Dauhauer, S.C. (2005*). Music as Therapy*. Southern Medical Journal, 98, pp. 282 -288.

Koelsch, S. (2010.) *Towards a Neural Basis of Music – Evoked Emotions*. Trends in Cognitive Sciensces, 1483), pp. 131-137.

Kurtović, A. (2013). *Odnos perfekcionizma i socijalne podrške s anksioznošću i depresivnošću kod studenata* [The Relationship of Perfectionism and Social Support with Anxiety and Depression in University Students]. Medica Jadertina, 43(4), pp. 189-200.

Mazo, E. (2002). *The Medicine of Music: Can music heal you?* Health, 6, pp. 74-81.

McCarthy, S. (2016). *The Therapeutic Power of Music: A literature Review*. Journal of the Australian Traditional-Medicine Society, 22(3), pp. 154-160.

Milivojević, Z. (2010). *Emocije - Psihoterapija i razumijevanje emocija* [Emotions – Psychotherapy and understanding of Emotions]. Zagreb: Mozaik knjiga.

Nadarević, T., Štefanac Nadarević, V. (2017). *Asklepiju i Orfeju u čast – deset godina aktivnosti na popularizaciji veza medicine i glazbe* [Asclepius and Orpheus in Onor – Ten Years of Popularization of Medicine and Music Connections]. Acta Medico-Historica Adriatica, 15(1), pp 37-44.

Naulinger, J. (1988). *Psihologija dokolice* [Psychology of Free Time]. In: Zbornik Trećeg programa Radio Zagreba, pp. 21-22.

Petz, B. (2005). *Psihologijski rječnik* [Psychological Dictionary]. Jastrebarsko: Naklada Slap.

Previšić, V. (2000). *Slobodno vrijeme između pedagogijske teorije i odgojne prakse* [Free Time between Pedagogical Theory and Educational Practice]. Napredak, 141(4), pp. 403-410.

Raedler, L. A. (2014). *The Dynamic Partnership of Music and Medicine*. The Oncology Nurse, 7(2), pp. 18-19.

Ruskin, H., Spector, C. (2002). *Implementation of Serious Leisure as part of Leisure education in Israel*. Journal for Culture, Society and Education, pp. 106-134.

Sullivan, J.M. (2007). *Music for the Injured Soldier: A Contribution of American Women's Military Bands During WW2*. Journal of Music Therapy, 44 (3), pp. 282-305.

Tagg, P. (2001). *Popular Music Studies – Bridge or Barrier? Musica come ponte tra i popoli* (ed. Tonini, G.). Bolzano: Libreria Musicale Italiana.

Vidulin, S. (2018). *Playing Instrument in Regular Music Class: Musical-pedagogical and Didactics Implications*. In: *Music in Society* (ed. Bosnić, A., Hukić, N). Sarajevo: Musicological Society of the Federation of Bosnia and Herzegovina, Academy of Music, University of Sarajevo, pp. 147-164.

The Top Mental Health Challenges Facing Students www.bestcolleges.com accessed: 5/10/2019

www.audiocura.com/the-impact-of-music-on-neurochemistry/ (2013)

https://www.encyclopedia.com/people/medicine/medicine-biographies/bernardini-ramazzini (5/18/2019)

http://hbl.lzmk.hr/clanak.aspx?id=1057 (5/18/2019)

Wheeler, B. L. (2015). *Music therapy handbook.* New York, NY: Guilford Publications.

The Cross-Cultural Experiences of Music Therapists Working with Adolescent Girls with Trauma in Jamaica: A Phenomenological Inquiry

Lillie Klein

Abstract

The purpose of this phenomenological inquiry was to explore the cross-cultural experiences of music therapists working with adolescent girls with trauma in Jamaica. Two board-certified music therapists (MT-BC), who had previously worked within this setting and population, shared their experiences during individual 60-minute in-person audio-recorded interviews. A detailed synthesis of the data was conducted through thematic analysis. Significant unifying components emerged and were grouped into four essential categories: (a) reactions to the Jamaican context; (b) challenges conducting music therapy; (c) connections through music; and (d) impact on the music therapist. Results concluded that when working cross-culturally with adolescent girls with trauma in Jamaica, music therapists may experience culture shock and vicarious trauma. Music therapists may also experience music in a unifying way and gain new insight both professionally and personally. The overarching goal of this study is to serve as a career related educational resource that can enrich current and future therapists toward a wider understanding of cross-cultural music therapy and trauma care both in native and nonnative communities.

Introduction

This study explores the cross-cultural experiences of two board-certified music therapists who had worked in Jamaica with adolescent girls that were exposed to varying degrees of significant trauma. It is my belief that to better understand how to practice music therapy in foreign countries in a culturally sensitive way, especially when working with clients in different regions who have been exposed to traumatic material; we must first explore the unique factors that play a role in the music therapists' experience. In this section, I will underscore my personal experience with trauma and my professional work in Jamaica. I will then outline the significance and need for continued research towards this specific phenomena.

Personal Motivation

My desire to explore the cross-cultural experiences of music therapists working with adolescent girls with trauma in Jamaica was shaped by several prominent experiences. My initial interest in trauma care stems from a personal encounter with a natural disaster. On October 29th 2012, Superstorm Sandy pummeled the East Coast of the United States, causing hundreds of deaths and leaving thousands homeless (American Red Cross, 2018). That day my world was unexpectedly shaken as an estimated fourteen-foot surge barreled through my waterfront apartment, taking with it my personal mementos, my home, and my job, as well as my sense of safety and security. The massive toll of destruction left my community shocked and devastated. Exposure to traumatic events can abruptly alter one's perspective of the world, sense of self and functional ability, often causing both direct and indirect lasting effects for individuals and communities (Loewy & Hara, 2002; Stewart, 2010; Sutton, 2002; Woodward, 2015).

During the long and arduous restoration process, my neighbors and I began singing together; our harmonious songs seemed to provide a sense of comfort during the adverse time. Through music, we transformed from a devastated community into a united chorus. This shared experience was personally monumental in exhibiting music's ability to unify and aid during crises and was instrumental in shaping my desire to enter the field of music therapy.

In 2014, I began my graduate studies at Molloy College. During that year, I had the fortunate opportunity to travel as a volunteer music therapy graduate student to the country of Jamaica for the first time through the Jamaica Field Service Project (JAFSP). The JAFSP is a volunteer service-learning program that brings music therapy students and professionals to Jamaica to provide services throughout parts of the island. I explored the coastal reef, plunged off cliffs into the Caribbean crystal clear water, and hiked through the sacred mountains of Accompong. As I relished in the beauty of the island, I also observed firsthand the disparity throughout the rural countryside as we passed remote villages. I was captivated by the antithetical landscape, where enormous tropical resort oases resided amidst impoverished towns. Limited human resources, shortage of health care providers, and nominal funding illustrated a portrait of a neglected region. This practicum provided me a firsthand account of the Caribbean island as well as a cultivated awareness of how different cultural contexts can offer diverse experiences within music and health care. It afforded me the opportunity to broaden my clinical skills, while widening my understanding of how music therapy can serve international communities in need.

This insight would further serve me the following year, when I revisited Jamaica; this time working with a different music therapy program, organized by Physicians, Residents, and Nurses (PRN) Relief International, affiliated with Molloy College and Yale-New Haven Hospital. As an interdisciplinary global response medical team, members of PRN Relief International work closely with low-resource communities to deliver clinical care and promote education and health. Twice a year, professionals and students across a wide range of disciplines deliver essential care for various communities in Jamaica. As a supervised student, I was fortunate to be a part of the program's mental health care team, providing music therapy services for the first time to a safe-house facility for adolescent Jamaican girls. The state-operated safe-house facility serves as a transitional home for at-risk adolescent girls, ages twelve to eighteen. The facility was originally established to create a stable, supportive, and nurturing environment for girls who had experienced various types of traumatic events in their lives, such as domestic violence, sex trafficking, conflict with the law, sexual abuse and/or rape, and deemed in need of care and protection (Jamaican Child Protection and Family Services Agency [CPFS], 2018).

The following fall, I returned to Jamaica for a third time as a student, continuing to provide music therapy services for the girls residing at the safe-house, this time with a sense of familiarity, but under new supervision. Taken together, these experiences revealed to me a deep curiosity to understand the interplay between music therapy, cross-cultural care, and trauma care. As a newly certified professional in the field of music therapy, I consider these personal experiences immensely significant in helping frame my cultural perspective toward health and wellness.

Significance

Scholars have long worked to unveil the underlying processes cloaked beneath the traumatic experience (Loewy & Hara, 2002; Stewart, 2010; Sutton 2002; Woodward, 2015). Recent literature suggests that culture plays a major role in how we experience traumatic events and in ways that may amplify the impact. Therefore, practitioners must understand the cultural factors that shape the client's worldview. This vantage may provide a lens through which we can help the client frame and locate meaning within the trauma. I believe that implementing relevant and culturally sensitive care for international traumatized clients relies first on understanding cross-cultural factors the therapist might encounter. An exploration of how music therapists are working in this forum may help identify unique components that impact the therapeutic process, including potential challenges and risks, and personal reactions, as well as methods and goals employed. Therefore, this study is relevant for underscoring how music therapy, cross-cultural care, and trauma care are woven together. The context of this study, Jamaica, represents an ideal setting, both for its vibrant musical history and high need for trauma informed care. Thus, examining the music therapists' experience in this cultural context is significant toward gaining deeper insight and contributing to the field of music therapy and the clients we serve.

Need for the Study

Across the globe, an escalation in a wide range of traumatic events, both natural and human-made are leaving diverse populations increasingly vulnerable to traumatic exposure (CPFS, 2018; Samms & Cholewa, 2014). As a result, more music therapists can be seen working in nonnative countries. Although in its pioneering stages, the last few decades have seen the development of increasing interest in cross-cultural care, with several documented accounts concerning the music therapists' cross-cultural experience (Fitzsimons, 2016; Gadberry, 2014; Kim, 2008; Valentino, 2006; Woodward, 2015). Correspondingly, numerous studies highlight the role of music therapy in international trauma care (Bolger, 2012; Coombes, 2011; Lang & Mclnerney, 2002; Navarro Wagner, 2013, 2015; Pavlicevic, 2004; Stewart, 2010; Sutton, 2002; Woodward, 2015). However, these two inter-related perspectives are scarcely discussed in conjunction. Due to the void in research, it may be concluded that cross-cultural trauma informed care has yet to be fully developed in the field of music therapy. Fortunately, much can be inferred and supplemented from related fields (deVries, 1996; Wilson, 2013).

While the disparate research above suggests foreign settings may be characterized by unique challenges and circumstances for music therapy, there remains too little supporting research. Therefore, more exploration is needed to generate a clearer understanding of how both music therapy and trauma care vary in cross-cultural settings. To ensure culturally sensitive care, it is imperative that music therapists have access to relevant information regarding cross-cultural considerations likely to inform their experiences and work. Further insight will not only encourage our field to explore diverse ways of practice, but inform practitioners both in native and nonnative communities. To my knowledge, no studies currently exist in regard to music therapy and the country of Jamaica. With statistics suggesting a high and rising prevalence of risk for exposure to traumatic material in the country of Jamaica, especially with female adolescents, there is growing need for applicable trauma informed services (CPFS, 2018; Samms & Cholewa, 2014). With limited resources on the island and a deep-rooted cultural stigma toward mental health care, alternative modalities of treatment such as music therapy are especially promising.

Literature Review

In this section, I will offer an examination of relevant literature including: (a) cross-cultural care; (b) music therapy and cross-cultural care; (c) trauma care; (d) music therapy and trauma care; and (e) cross-cultural music therapy and trauma care. This will be followed by a brief discussion about the context of Jamaica: (a) history of Jamaica; (b) music of Jamaica; (c) trauma concerns in Jamaica; (d) mental health care perspectives in Jamaica; and (e) music therapy in Jamaica. Reference material has been acquired through Molloy College supported databases and includes recent peer-reviewed articles and seminal works both in music therapy and relevant fields. Published books have been retrieved to obtain additional theoretical perspectives. The majority of the literature reviewed is from the year 2000 onward, although earlier sources were referred to in order to generate a contextual and historical understanding.

Cross-Cultural Care

At many levels, culture frames who we are and how we identify in the world. Culture influences our perceptions, interpretations, and modes of expression. Within the therapeutic relationship, the therapist and client both bring with them a unique cultural identity and varied life experiences. Therefore, understanding the cross-cultural interplay between the therapist and client is crucial when providing therapeutic services in a foreign country.

Defining culture. Early formulations of cultural studies referred to culture as "that complex whole which includes knowledge, belief, art, morals, laws, custom, and any other capabilities and habits acquired by man as a member of society" (Tylor, 1871, p. 1). Today the term "culture" takes on many forms in the literature and is often heard in relation to concepts such as geographical area, nation, race, ethnicity, language, norms, identity, and gender. Due to the varied interpretations, this study will delineate culture using Kohl's (1984) commonly accepted definition:

> An evolving, integrated system of learned behavior patterns that is characteristic of the members of any given society. Culture refers to the total way of life for a particular group of people. It includes what a group of people thinks, says, does and makes - its customs, language, material artifacts and shared systems of attitudes and feelings. Culture is learned and transmitted from generation to generation (p. 17).

Cultural systems and values are extremely significant in understanding how individuals behave and communicate (Bigby, 2003). According to Luckman (2000), culture helps to shape one's worldview. Additionally, culture provides us with perspective toward health and illness. While scholars have been examining the concept of culture for centuries, research has devoted significant attention to cross-cultural studies only over the past few decades (Bigby, 2003; Luckman, 2000). The terms "cross-cultural" and "multicultural" have been used interchangeably throughout the literature; although they have a slightly different implied meaning (Kim, 2008, p. 3). Unlike multiculturalism, which refers to a society that contains several different cultural, ethnic, or racial groups (Stige, 2002); the term cross-cultural specifically refers to comparing two visible cultural, ethnic, or racial groups, and understanding their values, beliefs, and norms (Kim, 2008). For such a demanding and complex area of work, cross-cultural care or cross-cultural therapy can be broadly defined as the ability of practitioners to provide effective services to two or more culturally diverse client populations (Bigby, 2003). The art of cross-cultural care concerns learning how to transcend one's own culture in order to form a therapeutic connection with clients from another culture.

Contributing factors to cross-cultural care. Studies show several influential qualities of cross-cultural care that may contribute to the therapeutic relationship, such as self-awareness, cultural empathy, worldview, knowledge of cultural norms and/or deviations, and a sense of cultural connection (Kim, 2008; Pavlicevic, 2004; Stige, 2002; Wheeler & Baker, 2010; Woodward, 2015). According to Kim (2008), personal awareness of cultural beliefs and background is a key element for understanding the dynamics of cultural relations and thus cross-cultural care. Stige (2002) offers a complementary perspective, expressing the need of cultural empathy or sensitivity when working with different cultural populations and suggests that therapists cannot truly understand others unless they understand themselves. Similar studies stress the importance of understanding one's worldview as way of locating meaning and purpose in the world (Kim, 2008; Pavlicevic, 2004; Stige, 2002; Wheeler & Baker, 2010; Woodward, 2015).

A person's worldview includes their various values, belief systems, and biases that have been developed as a result of cultural conditioning. Pavlicevic (2004) recommends looking at the whole person, including their worldview and their way of constructing meaning in the world. According to Kim (2008), if culture influences the way we see our worldview, and ourselves then "clearly culture must come into play as we assess, diagnose, plan treatment, and engage with our clients" (p. 1). Lastly, the concept of connection is shown to be a significant quality in cultural work (Wheeler & Baker, 2010; Woodward, 2015). Given the above, the shared experience of music making may also be understood as a cultural engagement, a way of building connection. Stige (2002) concludes that culture influences music therapy even if the client and therapist share certain aspects of their culture. Therefore, understanding the music of a culture may offer additional insight towards cross-cultural perspective.

Music Therapy and Cross-Cultural Care

In the book Defining Music Therapy, Bruscia (1998) interprets music therapy as "a systematic process of intervention wherein the therapist helps the client to promote health, using music experiences and the relationships that develop through them as dynamic forces of change" (p. 20). According to Bruscia (2002), music therapy practice evolved through forces of thought on how music influences human behavior and the physical world, unearthed unconscious dimensions, self-actualization, and spiritual development. Bruscia labeled culture-centeredness as the "fifth force in music therapy" (p. xv). Although music can be seen in many cultures and is often referred to as a universal phenomenon, anecdotal reports specifically suggest that people's perception of the qualities within music, such as rhythm, melody, and form, may not be universally shared (Ruud, 1998). Rather, these musical elements are associated more with culture and tradition. According to Stige (2002), music can be seen as an artifact of culture, stating that:

> Artifacts are the products of cultural processes, and both physical objects and meaning systems such as languages and musics are understood as artifacts. Artifacts are passed down from generation to generation, often with novel enhancements made to them...Over time artifacts evolve in complexity, form and/or functionality (p. 327).

This statement appears to align with Kohl's (1984) definition of the characteristics of culture.

Although we often see reference to the terms "culture-specific music therapy" and "culture-centered music therapy" throughout the literature, cross-cultural music therapy is not specifically defined (Stige, 2002, p. 42). According to Stige, culture-specific music therapy is the recognition that the music therapist and client both come from a culture and therefore, one's culture can influence therapeutic processes. Correspondingly, culture-centered music therapy is defined as a practice that grows "out of interests of action and knowledge situated in specific cultural and social circumstances" (p. 191). In regard to this study, cross-cultural music therapy is defined as the practice within the therapeutic relationship that involves two different cultural backgrounds and ethnicities (Kim, 2008). This aligns well with culture-centered music therapy as the cultural settings and social circumstances are embedded in the context where therapy occurs.

Approaches to cross-cultural music therapy. The field of music therapy continues to expand globally (American Music Therapy Association [AMTA], 2018a). According to the World Federation of Music Therapy (2018), eighteen countries have developed a national music therapy association. The AMTA (2018a) reports that five percent of members are living abroad in thirty international countries. In fact, the literature reveals that music therapists are rapidly increasing delivery of services abroad in foreign countries where music therapy is not widely practiced or where there is an absence of native trained and educated therapists (Bolger, 2012; Coombes, 2011; Lang & McInerney, 2002; Navarro Wagner, 2013, 2015; Pavlicevic, 2004; Stewart, 2010; Sutton, 2002; Woodward, 2015). Despite this, research on cross-cultural music therapy specifically appears to still be limited. However, in recent years a few pioneering music therapists have begun to extensively research the important role and interplay of the cross-cultural relationship in music therapy (Fitzsimons, 2016; Gadberry, 2014; Kim, 2008; Valentino, 2006; Woodward, 2015).

Considerations in cross-cultural music therapy. Previous accounts suggest that practicing music therapy in nonnative countries can be both rewarding and challenging, with a myriad of obstacles and barriers to manage (Bolger, 2012; Coombes, 2011; Lang & McInerney, 2002; Navarro Wagner, 2013, 2015; Pavlicevic, 2004; Woodward, 2015). For instance, music therapists practicing in foreign countries remarked that they felt differential dynamics in power and status or, at times, felt de-skilled (Bolger, 2012; Coombes, 2011; Pavlicevic, 2004; Woodward, 2015). These studies further revealed heightened feelings of being a foreigner, questioning therapeutic role, timetable, and boundary concerns. According to Bolger (2012), maintaining boundaries, protecting the physical space, and client privacy were seen as significant obstacles in music therapy sessions. In addition, language barriers were cited as a unique challenge, although this consideration may differ depending on the therapist's theoretical orientation (Lang & McInerney, 2002; Navarro Wagner, 2013, 2015). Navarro Wagner (2015) conveyed that although language is important to create bridges, it does not protrude as a "barrier for musicking" (para. 35). Alternatively, music therapists who use verbal techniques outside of music experiences remarked on the beneficial impact of having a translator co-facilitating amongst sessions (Lang & McInerney, 2002).

Literature further notes that music therapists may experience a sense of culture shock when working in a different country (Woodward, 2015). Culture shock is defined as a feeling of confusion or disorientation that occurs when an individual leaves a familiar place and shifts to an unfamiliar one (Levine & Adelman, 1993). Culture shock is said to be brought on by the anxiety of losing many familiar signs and symbols of social interaction. Social cues such as language, gestures, customs, or norms are all ways in which we orient ourselves to different situations in life. Perceptions of illness, social and economic circumstances, views on disability and health care provision can all contribute toward this initial shock. Levine and Adelman identified six stages of culture shock: (a) anxiety about leaving home; (b) new excitement about surroundings; (c) overwhelmed by the new environment; (d) adjustment stage; (e) loneliness, loss of self-esteem and setbacks; and (f) find balance and acceptance. Interestingly, Valentino (2006) revealed that working in a nonnative culture can affect every therapeutic encounter and cause the therapist to repeatedly consider cultural implications potentially more than their colleagues who work within their native country.

The literature suggests several influential qualities and circumstances that contribute to and affect the music therapists' cross-cultural experience (Fitzsimons, 2016; Gadberry, 2014; Kim, 2008; Valentino, 2006; Woodward, 2015). These central considerations not only emphasize the importance of recognizing one's own worldview and cultural perspective, but acknowledging one's personal theoretical orientation and primary goal when facilitating therapeutic treatment. Assessing differing cultural norms and surveying cultural identity, such as values, beliefs, ideals, prejudices and privileges, may help toward comprehending how our personal assumptions are shaped (Fitzsimons, 2016; Gadberry, 2014; Kim, 2008; Valentino, 2006; Woodward, 2015). Understanding the role of culture may help illustrate a portrait through which one may locate meaning to the experience of life. As will be discussed in detail below, these cultural considerations are particularly important when dealing with traumatic life experiences.

Trauma

Exposure to a traumatic encounter can abruptly alter one's perspective of the world, sense of self and functional ability in a shocking and disruptive manner (Loewy & Hara, 2002). Discerning how traumatic life events affect individuals and groups is a complex and varied discussion. Much of this complexity comes from the varied history, treatment, and etiological classification of trauma.

Defining trauma. Derived from the Greek word for "wound," trauma theories were originally fashioned from studies with male soldiers and were attributed to psychological hurt initiated through accidents and war (van der Kolk, 2014). In its modern usage, the term "trauma" can be described as an emotional, psychological, and/or physiological shock or wound that causes damage, pain or suffering and its lingering effects (Loewy & Hara, 2002; Stewart, 2010; Sutton, 2002). The current Diagnostic and Statistical Manual of Mental Disorders (DSM-V) (5th ed.; DSM-5; American Psychiatric Association [APA], 2013), describes trauma as witnessing or experiencing directly or vicariously an actual or threatened death, serious injury or sexual violence, and includes a variety of stressors of varying magnitude, frequency, and duration. Knafo (2004) describes the concept of trauma as:

> The rupture of continuity in time, in relations and attachments, in perceptions of self and others, in basic assumptions about the world, in future expectations, and, above all, in ruptures in one's fabric of meaning and belief systems (p. 171).

Levine (1997) describes trauma in musical terms as being an interruption to one's rhythm or flow of energy. According to Levine, the body's internal awareness or "felt sense" may become immobilized when exposed to a traumatic event. Levine points out that for many trauma survivors, awareness of feelings and the emotions that underlie them are initially inseparably linked to the immobility response - terror, horror, rage, and helplessness. Porges' (2001) polyvagal theory looks at understanding how the nervous system responds to threat and trauma explaining, "The autonomic nervous system is linked to the affective experience, emotional expression, facial gestures, vocal communication and contingent social behavior" (p. 127). Porges states that our bodies limbic defense systems that organize and regulate fight, flight and freeze behaviors are altered when we have unresolved trauma in our lives. Although the experience of trauma can be interpreted from numerous perspectives, three aspects are regarded as central to the understanding of trauma: shock, wound, and a lasting effect (APA, 2013).

Contributing factors to trauma care. Trauma takes many forms and impacts all ages. Stressors that may qualify as traumatic can come from a myriad of sources such as direct or indirect exposure to conflict and war, death or witness of death, natural disaster, refugee status, terrorism, domestic assault, rape, sex trafficking, and physical and/or sexual abuse (APA, 2013). Although the term is broadly used, potential symptomatology of trauma is individually embodied in a wide variety of ways, at a range of impactful levels, and interpreted meanings.

Exposure to traumatic material can have significant lasting repercussions and present a range of physical, neurological, and psychological impacts, including increased risk for potential maladjustment, suicidal or self-injurious behaviors, anxiety and panic attacks, substance abuse, and low-self-esteem (APA, 2013; van der Kolk, 2014). In addition, dissociation, somatization and affect dysregulation can be seen as highly interrelated among those exposed to trauma (van der Kolk, 2014). Trauma related disorders are diagnostically distinguished within two categories. First, acute stress disorder relates to the immediate effects of a traumatic event (APA, 2013). The second category, post-traumatic stress disorder (PTSD), refers to individuals who have been exposed to extreme stress or trauma; the symptoms consist of re-experiencing the event, avoiding reminders, and hyperarousal, such as increased startle reaction, insomnia or irritability that persists for over a month. For example, combat soldiers who have been exposed to life-threatening experiences are susceptible to developing PTSD after returning home from the war front. However, PTSD is not exclusively experienced through war; any individual exposed to traumatic stressors is equally at risk for post-trauma adjustment problems (van der Kolk, 2014). The lingering effects of trauma and PTSD can produce extreme behaviors, such as withdrawal or detachment, helplessness, lack of concentration, irritability, aggression and sleep disturbance (APA, 2013).

Specific methods of trauma informed care have shown to be beneficial for combating these symptoms, including: cognitive behavioral therapy (CBT), eye movement desensitization and reprocessing (EMDR), somatic therapy; medication, and various psychotherapy techniques and interventions (van der Kolk, 2014). Among these approaches music therapy has shown to be especially promising (Loewy & Hara, 2002; Stewart 2010; Sutton, 2002).

Music Therapy and Trauma Care

"As trauma is an experience of life - so is music" (Borczon, Jampel & Langson, 2010, p. 55). According to Wilson (2013), exposure to a traumatic event can overwhelm the ordinary systems of care that give people a sense of control, connection and meaning. Expressive art modalities, such as music, go beyond the conceptual meaning captured in language and have shown to provide a nonverbal outlet for emotions. Researchers have documented the significant role music therapy can play in supporting individuals and communities in coping with trauma related symptoms, particularly by fostering resiliency, safety, and support (Loewy & Hara, 2002; Stewart, 2010; Sutton, 2002; Woodward, 2015). These studies have reported to help trauma survivors increase feelings of control and empowerment, and gain perspective in a unique way by obtaining insight through creative expression.

Music therapy and trauma have a long history that is intrinsically linked. In fact, music therapy has its genesis working with traumatized soldiers after World War II (Ainlay, 1948; AMTA, 2018b). Initiated by the Army's Reconditioning Program, community musicians were sent to hospitals to play for veterans suffering from the effects of war and boost morale (Ainlay, 1948). This program was the "first actual recognition given to music as a definite means to be used in a hospital to help the sick and injured" (p. 323). Today, current studies show that music therapists are working both in home and foreign countries with a wide array of client trauma resulting from terrorism, natural disasters, and sex trafficking, to name a few (Bolger, 2012; Coombes, 2011; Lang & McInerney, 2002; Loewy & Hara, 2002; Navarro Wagner, 2013, 2015; Pavlicevic, 2004; Stewart, 2010; Sutton, 2002; Woodward, 2015).

Approaches to music therapy and trauma care. In recent years there has been an emergence of music therapy publications detailing varied theoretical perspectives and methods employed with specific trauma populations and ages (Amir, 2004; Austin, 2001; Bensimon, Amir & Wolf, 2008; Bonny, 2002; Juan, 2016; Sutton & Backer, 2009). For instance, Austin (2004) incorporated the use of voice with specific music psychotherapy techniques, such as vocal toning and holding, in an effort to promote intra/interpersonal change for adults who experienced childhood trauma. Bonny (2002) employed Guided Imagery and Music (GIM) methods with Vietnam veterans, exhibiting music therapy GIM techniques to be a supportive treatment towards helping veterans' symptomatology of trauma. In another study, Bensimon, Amir, and Wolf (2008) explored the use of group drumming with soldiers experiencing PTSD. Their study revealed some reduction in PTSD symptoms and found drumming to increase a "sense of openness, togetherness, belonging, sharing, closeness, connectedness and intimacy, as well as achieving a non-intimidating access to traumatic memories, facilitating an outlet for rage and regaining a sense of self-control" (p. 1). Juan (2016) examined the role of music therapy with domestically abused women from various cultures. One key factor that emerged in the study was the importance of taking into account the trauma survivors' cultural and social variables. This study remarks on the therapist's ability to understand culture and its relationship to music and identity to implement competent treatment. Amir (2004) utilized Analytical Music Therapy improvisation techniques with sexually abused clients and claimed that this method was effective in providing an outlet to unearth unconscious past traumatic material. Sutton and Backer (2009) investigated similar applications of spontaneous music making. The researchers explored improvisational musical play with one traumatized young boy and one adult, and discovered the notion of silence in trauma care. Data revealed the participants' tendency to fill up emptiness and "walls of sound" in their music. Surveying the literature, it commonly appeared that the main populations for music therapy and trauma care were focused toward war combatants, refugees affected by conflict, and natural disasters (Bensimon, Amir & Wolf, 2008; Bonny, 2002), while significantly less attention was given to varied types of traumatic stressors, such as domestic and/or sexual abuse (Amir, 2004; Juan, 2016).

Considerations in music therapy and trauma care. Numerous studies have examined the personal challenges and impact on music therapy professionals and students providing support for clients impacted by trauma (Gao, O'Callaghan, Magill, Zhang, Zhang & Shi, 2013; Loewy & Hara, 2002). In one report, Gao et al. (2013) describe an extreme array of emotions, attachments, and a sense of pressure when providing music therapy to victims of a natural disaster. Literature in trauma informed care illustrates common variables that may affect the therapist when working with clients who have experienced significant trauma; such factors are commonly referred to as "vicarious trauma," or "secondary traumatic stress" (Finklestein, Stein, Greene, Bronstein & Solomon 2015; Knafo, 2004). These terms describe the effects on the therapist from working with traumatized persons, although despite their conceptual similarities, their emphases differ.

Secondary traumatic stress, often referred to as "compassion fatigue," describes the sudden adverse reactions people can have when helping survivors of trauma, while vicarious trauma is salient for therapists who work with survivors of violence, particularly sexual assault and incest (Pearlman & Saakvitne, 1995). Vicarious trauma may be distinguished as the process of change over time that results from witnessing or hearing about others' suffering and feeling committed or responsible to help them. Pearlman and Saakvitne defined vicarious trauma as the permanent "transformation in the inner experience of the therapist that comes about as a result of empathic engagement with clients' trauma material" (p. 31). The main symptoms are disturbances in the therapist's cognitive frame of reference, "identity, worldview, and spirituality...affect tolerance, fundamental psychological needs, deeply held beliefs about self and others, interpersonal relationships, internal imagery, and physical presence in the world" (p. 280). Loewy and Hara (2002) speak to this in their seminal work Caring for the Caregiver: The Use of Music and Music Therapy in Grief and Trauma, which explores the impact on the therapist and caregiver from treating individuals exposed to the traumatic events of September 11th. It is said that over time, these emerging concerns can lead to changes in the therapists' psychological, physical, and spiritual well-being. Finklestein et al. (2015) addressed symptoms of vicarious trauma among practitioners working in communities exposed to high levels of trauma. Study findings suggest that self-efficacy or confidence in one's professional ability to treat and work with clients is a significant protective factor regarding traumatization. The following section demonstrates how the aforementioned issues are further compounded when the therapy takes place in a cross-cultural context.

Cross-Cultural Music Therapy and Trauma Care

Culture frames who we are and how we identify in the world, although the complexities of culture are expanded and exacerbated by the prevalence of trauma and need for trauma care. As stated earlier, culture affects how we cope and experience traumatic events and therefore, the role of culture may help to provide a lens through which one may frame and locate meaning within the traumatic experience (deVries, 1996; Wilson, 2013). Literature illustrates the critical need for mental health care professionals to broaden their understanding and treatment of trauma within a cultural context. In recent years, music therapists have begun to research the impact and important role cross-cultural care plays in music therapy (Fitzsimons, 2016; Gadberry, 2014; Kim, 2008; Valentino, 2006; Woodward, 2015). Correspondingly, numerous studies highlight the role of music therapy and international trauma care (Bolger, 2012; Coombes, 2011; Lang & McInerney, 2002; Navarro Wagner, 2013, 2015; Pavlicevic, 2004; Stewart, 2010; Sutton, 2002; Woodward, 2015). However, the conjoined significance of the two components of cross-cultural care and trauma care have been minimally articulated within the existing body of literature.

Approaches to cross-cultural music therapy and trauma care. Cultural perspective suggests that it is crucial for one to understand traumatic events within a broader context (Stige, 2002). However, documented data reports of music therapy and trauma care in foreign countries have primarily focused on the client's cross-cultural experience, with limited inquiry toward the music therapist's cross-cultural experience. Indeed, the existing body of literature tends to focus solely on ethnicity and language, while the role of culture is much more complex, specifically when related to trauma (deVries, 1996; Wilson, 2013).

Considerations in cross-cultural music therapy and trauma care. According to Wilson (2013), to better understand and help treat trauma survivors, the therapist must first comprehend the sociocultural environment in which the trauma exists. Correspondingly, deVries (1996) illustrates the vital role culture plays in individuals' ability to cope with potentially traumatizing experiences by creating "meaning systems" that explain the causes of traumatic events. In effect, deVries suggests that culture can act as a filter toward the perception of trauma. Trauma responses should therefore be respectfully understood within a cultural context.

Correspondingly, Stige (2002) suggests that "meaning systems" are embedded with music, and can be seen as a cultural artifact that contains various interpretations and roles in different cultural contexts. However, related literature has seldom evaluated the totality of these considerations. It appears that additional research documenting the cross-cultural experiences of music therapists working in foreign countries with client trauma is needed (Woodward, 2015). Future studies may serve to provide a better understanding of the interplay between music therapy, culture, and trauma care. To this end, the current study addresses this gap in the literature by investigating music therapists' experiences in Jamaica, a context ideally suited for looking at the intersection of music therapy, cross-cultural care, and trauma care. The next section elucidates this context.

The Context of Jamaica

Jamaica is the third largest island of the Greater Antilles in the Caribbean. The tropical country is divided into fourteen parishes, with its capital being Kingston, located on the southeast coast (Black, 1994). Jamaica's estimated population is two and a half million (Mason, 2000). Economically, Jamaica is classified as a developing country, with around fourteen percent of the population living below the poverty line. Mountains, beaches, coastal plains, and rain forests encompass the sun-drenched tropical climate. The natural riches and scenic beauty of Jamaica serve as the main driver for tourism, which accounts for sixty percent of the domestic product, while music is arguably Jamaica's most recognizable export.

The Jamaican national motto, "Out of Many, One People," speaks to diversity and unity regardless of differences. The motto is based upon the varied cultural backgrounds of the people who historically settled on the Caribbean island (Mason, 2000). Today, the motto has come to symbolize the embodiment of national pride and the mixture of cultures that give Jamaica its unique identity. This complex blend is further characterized by rich cultural traditions and tempered by political instability, financial indebtedness, social inequality, and revolution. Although these factors contribute to one's understanding of Jamaica, they cannot encapsulate the multidimensional perspectives of the lush tropical island and its people. This section offers a brief overview of Jamaica. The role of this literature review is not to shape biases or assumptions, but merely to inform and situate the reader toward potential cultural considerations.

History of Jamaica. Knowing the historical context of the Jamaican heritage is crucial toward understanding the dynamics of its present state and conditions. Jamaica's original inhabitants were said to be the Arawaks (Black, 1994). They named the island Xaymaca, meaning land of wood and water. The indigenous Arawak community was decimated by slavery, war, and disease brought by the Spanish in 1494 when Christopher Columbus discovered the island and colonized. The Spanish brought the first slaves from Africa to work on the plantations, mainly cropping sugarcane. The

English later invaded and took control of the island in 1655. During this time, many of the Spanish slaves escaped and fled to the mountains, later to be known as the Maroons. The colonization of the island resulted in a large importation of slaves from West African regions. Slavery in Jamaica did not end until 1808, when the Abolition Bill was passed. In 1962, Jamaica gained its political independence from British rule, but still remains a Commonwealth nation. Jamaican culture is a vibrant and proudly independent democratic society. English is Jamaica's official language, although it is spoken with many variations ranging from British English to Jamaican Patois. Patois is labeled as a Jamaican Creole dialect, which evolved from various linguistic backgrounds. It is a melodic, rhythmic language, that follows its own grammatical rules and logic, which preserves the ethnic and cultural identity of Jamaica.

Music of Jamaica. Music is said to be a central feature of the social and cultural life of Jamaicans (Lewin, 2000). In Jamaica, music assumes many roles, including a mode of expression of basic philosophies, attitudes, beliefs, a vital link with the spiritual world, and a cultural connector. According to Sutton (2002), music profoundly shapes the goals and objectives of a people moving toward collective identity, cultural nationalism, and political independence. Therefore, to understand Jamaica's music, one must first understand the total environment of the people who create it. Historically, music has played a functional and significant role in Jamaica and can be heard throughout religious rituals, ceremonies, and all stages of life (Lewin, 2000). A review of Jamaica's music revealed layered influences, including remnants of sounds of the indigenous people, European colonization, and a variety of African rhythms (Davis & Simon, 1992; Lewin, 2000). This multicultural collision of Afro-Caribbean characteristics and British features contributed to the island's musical development.

Jamaica is considered the most Africanized of the Caribbean islands; this can be seen in the traditional dancing and heard in the native drumming rhythms (Lewin, 2000). Some of Jamaica's deep-rooted musical traditions are preserved within religious practice and ritual ceremonies. For example, group improvisatory drumming is perceived as sacred for some individuals in Jamaica and is often played for hours and said to induce trance states and facilitate direct communication with the spirit world. Drumming and percussive instrumentation is widely used in all levels of community activities and rarely exists without dance and movement. As Stowell (2000) describes:

> The functionality of music couples with a theme of collective participation. The audience is active and essential to the music. The music does not exist without significant involvement of work, dance, song or clapping of hands (para. 10).

Another key musical component is the function of the beat. The beat is an essential part of Jamaican music, and the specific rhythmic components capture the spirit of the island. These rhythms encompass traditional mento or folk music, calypso, and soca styles, as well as contemporary rocksteady, ska, dancehall, and reggae music (Lewin, 2000). Mento music is considered Jamaican folk music that evolved from chants and songs the slaves would sang while working on the plantations (Davis & Simon, 1992). These songs were adapted from traditional British work songs with a shuffle rhythm and can still be heard in Jamaican children's songs (Lewin, 2000). Ska and rocksteady music emerged around the same time Jamaica was fighting for independence from Britain (Davis & Simon, 1992). This music was heavily influenced by Western society, with a focus on removing beats from the shuffle rhythm, giving Jamaican music its distinctive quality of accenting the offbeat. This style of music would eventually shift into what is known today as reggae music.

Reggae is Jamaica's national music and a recognizable cultural asset (Davis & Simon, 1992). Many Jamaican musical styles have contributed to the reggae sound, but it is primarily a synthesis of American rhythm and blues, blended with traditional mento music. Reggae is Jamaican soul music, with accents on the second and fourth beats, with an emphasis on the bass as the principal melodic drive of most traditional songs. Lewin (2000) exclaimed, "If you want to hear the heartbeat of the Jamaican people listen to reggae music" (p. 26). It is said to have originated in the city of Kingston, although no one knows for certain where the word came from (Davis &

Simon, 1992). Some historians trace the term "reggae" to the Jamaican dialect word for "raggedness" and claim it means "coming from the people." The word was said to have first appeared on the 1967 record by the Maytals called Do the Reggay. Since then it has become a worldwide phenomenon, inspiring musicians from various cultures. Reggae continues to develop both traditionally and with strong hip-hop influences, creating dancehall music, a mixture of disc jockey culture and pop music (Lewin, 2000). Notable reggae artists include Desmond Dekker, Peter Tosh, Jimmy Cliff, Toots and the Maytals, and Bob Marley and the Wailers (Davis & Simon, 1992).

Considered to be the most famous reggae innovator and a national hero, Bob Marley introduced the genre to an international stage and gave Jamaicans a global voice (Lewin, 2000). Bob Marley used music as a tool to address political unrest, while promoting equality, and popularizing spiritual concepts of Jamaica and the Rastafarian religion. The Rastafarians are members of Jamaica's spiritual nationality with deep political convictions (Davis & Simon, 1992). Music holds an important spiritual, political, and emotional power for this religion. The basic percussive Rasta-rhythm is a sustained two-beat riff that mimics the human heartbeat and is believed to reach the human soul. According to Lewin (2000), reggae music is fueled by spirituality, political and socioeconomic commentary, as well as reflections of the Jamaican lifestyle and history of enslavement. In a society where the poor have few formal means of expression, reggae has given millions of Jamaicans the freedom to shape something in their own cultural identity.

Trauma concerns in Jamaica. Various socioeconomic vulnerabilities, such as poverty and social violence, greatly affect the current Jamaican population (Mason, 2000). These factors fuel a wide range of critical social issues that leave many susceptible to the exposure of traumatic material. In recent years, Jamaica has been cited as one of the most violent countries to live in, with a significantly high murder rate. The high incidence of homicide and physical violence in Jamaica is largely attributed to domestic disputes, gang-related conflicts, political tribalism, and a breakdown in social order. In addition, sexual abuse, rape, incest, and sex trafficking are prevalent throughout Jamaican communities, and it is estimated that over fifty percent of reported cases target adolescent girls between the ages of ten and nineteen. From 2008 to 2012, there were over seven thousand reported cases of sexually abused adolescents in Jamaica (Samms & Cholewa, 2014). According to the Jamaica Adolescent Reproductive Health Project (2003), sexually abused individuals are more likely to be assaulted by persons known to them. Due to the large-scale impact of sexual abuse, specifically affecting young Jamaican females, there appears to be a growing demand for adolescent trauma informed mental health care services in Jamaica.

Mental health care perspectives in Jamaica. Historically, Jamaicans have been hesitant to seek mental health care (Palmer, Palmer & Payne, 2012). Instead, individuals have been known to pursue support from respected leaders in their communities, primarily relying on pastors, family members, and teachers. Studies have reported that adolescents in Jamaica typically look to friends and family for support, rather than health care professionals (Samms & Cholewa, 2014). According to Palmer et al. (2012), “Seeking outside help is regarded as sharing personal issues with a stranger, with the potential consequence of placing the family at risk of being stigmatized” (p. 99). Stigma toward mental illness in Jamaica can be seen throughout the literature, revealing perspectives that only “crazy” or “mad” individuals were in need of care. Correspondingly, mental illness in Jamaica is often seen through a religious or superstitious lens. It may be perceived that individuals affected by mental illness are being punished for bad behavior by a supernatural or higher power. It is important to consider these deep-rooted beliefs about mental health care and how they may present barriers for practitioners.

Music therapy in Jamaica. To my knowledge, there are no current published studies in regard to music therapy in Jamaica. Despite scholarship lagging, music therapy practice is present within the country. For instance, The JAFSP has offered music therapy services to various populations throughout the island through a volunteer music therapy program for several years. Recently, PRN Relief International has begun to provide music therapy services in Jamaica. However, each of these programs currently only implements short-term therapeutic care. Sources revealed that there is yet to be any established music therapy professionals permanently practicing in Jamaica. As illustrated by Davis and Simon (1992), music is significant to Jamaican culture. Therefore, although in its genesis in Jamaica, music therapy might serve as a harmonious and complementary form of psychotherapy. Music, being a familiar and nonthreatening modality within the Jamaican culture, may act as an ideal vehicle to overcome mental health care stigma and the complex issues facing the people of Jamaica.

Summary

The country of Jamaica, with its vibrant music and gentle tropical landscape, has had a turbulent history fueled with trauma. From the Spanish settlement to British rule, slavery to political independence, these factors are still affecting the country today and the many adversities its people face as a small developing democratic island. From this, researchers are calling for wider implementation of applicable trauma informed care, especially concerning female adolescents (CPFS, 2018; Samms & Cholewa, 2014). With limited resources on the island and a cultural stigma toward mental health care, alternative modalities of treatment such as music therapy should be investigated.

While there is a vast quantity of compelling literature that examines music therapy, cross-cultural care, and trauma care independently, published work that considers the specific interplay between these three factors remains underdeveloped. As a comprehensive global health field, it is important that music therapy scholarship develops deeper insights to help inform the field and the clients we serve. Further research is needed to better understand how music therapists may effectively implement relevant and culturally sensitive trauma informed care in foreign cultures and countries. This sentiment guides the research questions described below.

Research Questions

For this study, I interviewed music therapists who had prior experience working cross-culturally in Jamaica with adolescent girls with trauma. The following overarching research questions helped to guide this phenomenological inquiry: (a) What are the cross-cultural experiences of music therapists working with adolescent girls with trauma in Jamaica? (b) What cross-cultural considerations occur while working with this population? and (c) How are those experiences reflected in music experiences?

Method

This phenomenological inquiry employed qualitative research methods to yield insight into the cross-cultural experiences of music therapists working with adolescent girls with trauma in Jamaica. In this section, I will introduce the study's design and thematic analysis used to synthesize the data, and delineate the specific research questions that helped guide this study.

Study Design

Phenomenology is a qualitative approach toward analysis, whereby the researcher seeks to uncover meaning among people who have had a common lived experience (Creswell, 2014; Wheeler, 2016). As the study concerns a specific phenomenon not widely researched, phenomenological design methods were appropriated to correlate data. In this study, data was collected through open-ended semi-structured interview questions with two board-certified music therapists who had previously worked with adolescent girls with trauma in Jamaica and met outlined criteria for the study, described below. Individual 60-minute audio-recorded interviews were conducted in-person and later transcribed and analyzed using thematic analysis (Braun & Clarke, 2006). Key unifying elements relating to the research questions were extracted from each interview, and essential categories and their respective thematic findings were identified. Member checking was employed to enhance credibility. Member checking is a qualitative research technique, wherein participants read through summarized interviews to verify their experience was accurately portrayed (Creswell, 2014; Wheeler 2016).

Participants

The two participants were recruited using purposive sampling, which is a non-probability sampling technique in which each participant is selected based on characteristics of the population and the objective of the study (Creswell, 2014). To ensure qualified recruitment, eligible music therapists were invited to participate in the study based upon the following inclusion criteria: (a) board-certified music therapist, (b) experience working as a board-certified music therapist in Jamaica with adolescent girls with trauma, (c) comfortable and open toward sharing details about their experience, and (d) never held residency or citizenship in Jamaica.

After approval was granted from the Institutional Review Board (IRB) of Molloy College, board reference number: 12111205-1215 (see Appendix A), an invitational e-mail was distributed to perspective research participants with information regarding the purpose of the study and detailing the specific inclusion criteria for involvement (see Appendix B). Interested participants contacted the researcher directly to join the study by the predetermined response date. Once selected, informed consent was provided to participants via e-mail to complete (see Appendix C). The consent form included information about the purpose of the study, as well as information regarding confidentiality and data security. An additional consent form for permission to audio record was provided (see Appendix D).

During the interview, the two participants were asked to share general information about themselves through an open-ended semi-structured demographic questionnaire. This collected data offered a perspective toward where the participants culturally situate themselves. Music Therapist A and B briefly shared their personal, cultural, educational, theoretical, musical, and professional background, as well as their affiliation with Jamaica prior to the cross-cultural experience. Table 1 presents the two participants' demographics.

Table 1. Participant Demographics

Participant	Music Therapist A	Music Therapist B
Age	40's	20's
Gender	Female	Male
Country of Origin	United States of America	United States of America
Ethnicity	Caucasian	Caucasian/Hispanic
Education	MS in Special/Deaf Ed. Bachelors in Music Therapy MT-BC, LCAT	MS & Bachelors in Music Therapy MT-BC, LCAT-P Advanced Training: NRMT, Level 1
Professional Experience	25 Years	4 Years
Clinical Population Experience	Geriatrics Child and Adolescent Psychiatric Pediatric Medical (Neonatal Intensive Unit, Oncology, Surgical, Palliative and Hospice Care) Early Childhood Special Ed.	Developmentally Delayed Children and Young Adults Group Home Settings with Developmentally Disabled.
Orientation	Humanistic	Humanistic, Music-Centered
Musical Background	Musical Theatre Vocalist Clinical Guitar and Piano	Jazz and Classical Guitar A cappella Music Performance
Jamaican Cultural Experience	Never Traveled to Jamaica Prior No Jamaican Family/Friends Prior	Never Traveled to Jamaica Prior No Jamaican Family/Friends Prior

Music Therapist A.

Music Therapist A identified as a Caucasian female, in her forties, who was raised and resides in a metropolitan area in the United States. She is a board-certified music therapist, licensed creative arts therapist, and has been practicing for twenty-five years. She received her bachelor's degree in music therapy and her master's degree in special and deaf education. Music Therapist A has worked with various populations and settings, including geriatrics, child and adolescent psychiatric and early childhood special education, with a focus on Spanish-English as a second language and deaf education. Additional settings include pediatric medical units such as oncology, neonatal intensive unit, surgical, and palliative and hospice care. Music Therapist A also teaches college and graduate level music therapy students and identified as having a humanistic orientation toward clinical practice. Music Therapist A's music background included training as a vocalist and experience in musical theater, as well as clinical guitar and piano training through a music therapy degree. Prior to the cross-cultural experience, Music Therapist A reported that she had no family or close acquaintances of Jamaican heritage and had never traveled to the country. In addition, the cross-cultural experience was said to have been her first time practicing music therapy in a different country.

Music Therapist B.

Music Therapist B identified as a Caucasian/Hispanic male, in his twenties, who was raised and resides in a metropolitan area in the United States. He is a board-certified music therapist, licensed creative arts therapist permit pending, and has been practicing for four years. He received both his bachelor's and master's degrees in music therapy, with level one advanced training in Nordoff-Robbins Music Therapy (NRMT). Music Therapist B has worked with various populations and settings, including children and young adults with special needs. Music Therapist B also teaches college level music therapy students and identified as having a humanistic orientation and music-centered approach toward clinical practice. Music Therapist B's music background included jazz and classical guitar and a cappella music performance. Prior to the cross-cultural experience, Music Therapist B reported that he had no family or close acquaintances of Jamaican heritage and had never traveled to the country. In addition, the cross-cultural experience was said to have been his first time practicing music therapy in a different country.

Setting and Materials

The individual participants designated the date and location for the in-person interview based upon their convenience. Interviews were conducted in a quiet and private location, with minimal distractions. A digital audio recording device was used for the interview and all data was stored on a personal computer, secured and locked with a password.

Data Collection Procedures

Data was collected by way of two individual interviews that lasted no longer than 60-minutes. The first part of the interview consisted of a brief introduction, where the researcher reviewed the purpose of the study, confidentiality procedures, and encouraged participants to answer each question to the best of his or her ability. Next, the participants were asked to describe their cultural identity through an open-ended semi-structured demographic questionnaire (see Appendix E). Participants shared demographic information, such as their cultural, educational, theoretical, musical, and professional background, as well as their affiliation with Jamaica prior to the cross-cultural experience. This information was regarded as appropriate toward understanding the participant's cultural perspective and its potential influence on the data. The second part of the interview consisted of open-ended semi-structured questions, with careful consideration not to lead the participants in any specific direction. Interviews were later transcribed into a written document and participants were asked to review the individual transcript for accuracy and verification.

Data Analysis

Once data had been confirmed by all participants, the data was manually coded according to thematic analysis, generating emerging categories and themes based upon participant responses. According to Boyatzis, (1998), thematic analysis is a process of correlating information into an organized form that is easily communicated to others. Braun and Clarke (2006) explained the method as a flexible approach based on data rather than preconceptions, which can deliver meaningful interpretation of data with no predetermined theory. In this method, themes are made by identifying, analyzing, and reporting patterns in relation to the data and interpreting themes to extract meaning and understand the specific phenomenon. Study findings were conducted and analyzed through the following six phase thematic analysis adapted from Braun and Clarke: (a) familiarize self with the data and then transcribe interview recordings verbatim into a written document; (b) generate initial codes by noting items of interest; (c) organize items of interest into relating topical categories; (d) review and create table of themes; (e) define and create a provisional name for themes; and (f) transform analysis into written summary report, and extract examples that relate to the themes, research questions, and literature.

Confidentiality and Ethical Considerations

In order to maintain safety and confidentiality, the study at all times adhered to ethical standards of clinical practice. To ensure that personal information was kept confidential, individual participants were provided anonymous pseudonyms: "Music Therapist A" and "Music Therapist B." Pseudonyms were maintained throughout the study and data collection procedures. Printed copies of the data identified the participants only by their anonymous pseudonym. Audio-recorded interviews and any identifying information was not shared with thesis advisor or committee members. However, interview data was discussed in an effort to help facilitate an understanding of emerging themes.

Results

The purpose of this study was to examine the cross-cultural experiences of music therapists working with adolescent girls with trauma in Jamaica. Two participants voluntarily consented to take part in the study. Individual interviews were conducted in-person, audio-recorded, and participant responses were later transcribed verbatim. The data was then coded using thematic analysis. This section is based upon the participants' interview responses when asked to describe their experience. Thematic analysis revealed four prominent categories that appeared to shape the participants' cross-cultural experience and were organized as follows: (a) reactions to the Jamaican context; (b) challenges conducting music therapy; (c) connections through music; and (d) impact on the music therapist. Each category section summarizes thematic findings and significant material that was discovered through the data analysis. Table 2 represents each category and their respective thematic findings.

Table 2. Categories and Thematic Findings

Category	Themes
Reactions to the Jamaican Context	Culture shock Countertransference
Challenges Conducting Music Therapy	Non-therapeutic work environment Language barriers Short-term care Negotiating cultural stigma
Connections Through Music	Cultural appreciation Promotion of stability Unity
Impact on the Music Therapist	Vicarious trauma Transformative Widening worldview

Category One: Reactions to the Jamaican Context

While working as a music therapist in Jamaica, each participant reported feeling significant reactions toward the cultural environment and norms, which appeared to impact their particular cross-cultural experience. According to the interview data, two predominant themes were identified in regard to the participants' reactions: (a) culture shock; and (b) countertransference.

Culture shock

A dominant reaction participants reported was experiencing a sense of cultural shock, particularly when observing the disparities in the region's socioeconomic status. Music Therapist A described feeling taken aback after witnessing the stark juxtaposition between wealth and poverty, remarking,

"The contrast was just so pronounced...There's such poverty in the area, it just evoked a lot of emotion in me." A similar response was reported by Music Therapist B: "I don't think I have ever worked with anybody as marginalized, ever. I don't think I ever bore witness to such an account of extreme poverty." In addition, culture shock reactions were prevalent for each participant upon observing the conditions of the safe-house facility and its visual surroundings. Music Therapist A reported an elicitation of emotions and described the atmosphere as being "very different, it was surrounded by fencing and barbed wire, with locked gates even inside the building." Correspondingly, Music Therapist B expressed feeling shocked upon the initial sight of the safe-house:

> You got a sense of the facility, as soon as you were driving on the bus and came around the top of the hill. You could see that it was barbed wired in and that you had to be let in the front gate. Almost everything was just dirt, sometimes falling apart, and everything looks unkempt and broken.

Countertransference.

While working in the cultural environment, each participant acknowledged being highly aware of how his or her culture related countertransference manifested. Music Therapist A and B recounted never fully acclimating to the environment, but remarked on allowing their perceived feelings of culture related countertransference to inform the clinical work. Music Therapist A expressed, "I think it was shocking, the environment was shocking to me, and it took some getting used to, although, I think, in part the countertransference informed some of the work." Similarly, Music Therapist B noticed an "overwhelming sense of countertransference that would just jump into me, that I just had to work and understand, as I was getting to know the girls."

Category Two:

Challenges Conducting Music Therapy

This category describes varying challenges that emerged while working with the adolescent girls in Jamaica. It is summarized in four main themes: (a) non-therapeutic work environment; (b) language barriers; (c) short-term care; and (d) negotiating cultural stigma.

Non-therapeutic work environment. The safe-house facility in which music therapy took place was reported by each participant as being a challenging environment and not conducive for music therapy. As expressed by Music Therapist A, "We had to provide therapy in a space that was not therapeutic." Moreover, one prevailing factor emerged as the most significant to have impacted the conditions of the therapeutic space, the acoustics. Music Therapist A remarked, "It was locked, it was loud, it was dark visually and acoustically...That impacted some of the work...It all lent to not feeling like a therapeutic environment." Music Therapist B reaffirmed this perception:

> It was a loud environment, all the time, it's always like metal clanging and girls screaming and a staff member yelling at someone to get down or something going on, there's a slap, there's a rock being thrown, you can hear all of these, just loud and abrasive sounds all the time.

Language barriers. Other challenge participants reported were varying degrees of difficulty understanding the Jamaican Patois dialect and specific cultural terms and phrases. Music Therapist A indicated struggles understanding and recalling the girls' names and explained this as being an ongoing challenge: "I knew English was the primary language there, but didn't realize how the Patois would be infused and I was caught off guard at times when I wanted to be therapeutic and did not understand what the girls were saying." Music Therapist B experienced similar moments when he had "difficulty understanding the girls' names to some degree."

Short-term care. Each participant identified challenges regarding aspects of the limited period provided for care. Music Therapist A stated, "It was too short...Basically everyday was assessment, treatment, and evaluation, all continuous and ongoing." This description was closely mirrored by Music Therapist B, who remarked, "It was extremely brief...You had to do a lot of getting to know, observing, assessment, the work, and then you're out of there, termination. So, I was fortunate that we had a lot of clinical time every day." Due to the nature of short-term care, both participants expressed their wariness of how deep to go with the trauma care and therapeutic process. Music Therapist A described the careful nuances of providing a level of support in the timeframe provided and balancing what the client needed in-the-moment:

> How deep could we actually go with our work...We're talking about trauma work, obviously we don't want to continue to re-traumatize, especially if we have such a short period of time, but the moments do present themselves and you think, "I can't really avoid going deeper right now, she needs it or it's appropriate" and how to really balance that in the moment, so we're validating their feeling and their experience around the trauma without going so deep that we don't have time to continue addressing it.

Music Therapist B echoed this idea of balance in the work:

> We would have to work with these girls for a long time, if ever being able to have a re-corrective emotional experience about sexual trauma...But, what we can try to do is work within the trauma, to develop a sense of trust and have just, hopefully a little bit of healing through just understanding it.

Negotiating cultural stigma. Each participant acknowledged an awareness of culturally specific stigmas surrounding the trauma experienced by the adolescent girls. Correspondingly, participants reported observing "typical" behaviors embodied in adolescent trauma. Music Therapist A explained, "I think so much of the behaviors you see are typical teen behaviors or typical trauma-related responses, that maybe look a little different because of the culture...Closed off, in terms of their experience and their fear with trust." Music Therapist B also shared that the cultural "stigma, was mostly why all those girls were there, because of their background of sexual abuse," remarking, "They had no other resources, nobody else to trust and no one else to believe in them anymore, because a lot of them had been raped or sexually assaulted by a member of their family."

Interestingly, both participants addressed the notion of how trauma affects the clients' chronological age. For instance, Music Therapist A mentioned witnessing the adolescent girls having the opportunity to be childlike: "They were suddenly able to be children and participate in something new and sharing of themselves, and be vulnerable, and playful, and childlike." Music Therapist B reported a similar account:

> I also think in a defensive mechanism way, they act even more

> childlike than they probably are chronologically...You have these extremes all around their personalities, in terms of, sometimes they are acting like adults even though they are thirteen, and sometimes they are acting like they are five and they are seventeen. It's understandable because of the amount of trauma that has been their norm.

Due to the significant trauma and the cultural stigma attached to the trauma, participants stressed the importance of developing trust within the therapeutic relationship. Music Therapist A expressed that the girls "were craving new and trusting relationships." A comparable perspective was disclosed by Music Therapist B:

> Working with the stigma and the trauma was most, if not all of the work...So a big thing we were working with was developing a sense of trust within the trauma...As I started to work and understand the girls, I could see this hardness and shell that protects them in a place like this, start to chip away.

Category Three:

Connections through Music

This category is based upon the participants' cross-cultural experiences, in regard to the music shared with the girls and how the music was personally experienced for each participant while in Jamaica. The data findings uncovered three main themes: (a) cultural appreciation; (b) promotion of stability; and (c) unity.

Cultural appreciation. Concerning preparation for the mission trip to Jamaica, each participant indicated the importance of acquiring a traditional and current sense of Jamaican music, as well as comprehending music's cultural significance in Jamaica. Music Therapist A reported having benefited from developing a cultural repertoire: "I think it really helped bridge a gap between our two cultures and really develop that sense of connection and support." Building an awareness of Jamaican music and culture prior to the clinical work was equally beneficial for Music Therapist B: "I tried to understand the cultural popular references, get an idea of what music is like down there, and familiarize myself with traditional Jamaican music, and percussion, and rhythms." In addition, the two participants each remarked that music appeared to be deeply significant for the adolescent Jamaican girls, specifically in regard to drumming and chanting. Music Therapist A recalled a specific instance witnessing how music was embedded within the culture:

> Drumming brought out rhythms and movement that manifested in a very specific cultural way...The music making in that moment, was not just sitting and playing rhythms on a drum. It was complete. It was like the music, the drumming, was infused in their soul in that moment and in a way where the music, they just became the music.

Music Therapist B additionally reflected that numerous music therapy experiences appeared to be relevant and culturally symbolic:

> We fully explored the gamut, in terms of working within the methods of music therapy. I felt like they all had their time and place and function...Sometimes we would get into purely improvisational either drumming or instrumental play, improvising a chant...Just being in a musical experience was gratifying, especially because music is so important to the culture.

Promotion of stability. Each participant reported to have utilized the qualities of music and music experiences to temper erupting moments of chaos within the environment. Music Therapist A shared how the escalating turbulence was often alleviated through the use of music: "The music was able to sort of reign in this aggressive chaos...Whether in the chant, a rain stick, a drum or one of our voices. Almost every time, turned the aggressive chaos to this beautiful shared moment." Music Therapist B as well felt that he was "jumping into a little bit of chaos," and found that through the use of music the disarray would dissipate:

> Music would often be used as a diffuser, if a fight was erupting or someone was having a tough time being there completely, we would use music as a way of diffusing the situation and try to find some sort of stability.

Unity. Each participant remarked on witnessing moments when spontaneous music making would emerge as a supportive and unifying resource, particularly during intense emotions or conflict. Music Therapist A shared a moment she observed when "all of a sudden one of the girls would start the chant or all of a sudden one of the girls would begin to sing our theme song, and found one or both of those things to be grounding, and calming, and foster cohesion." Recalling a specific moment while working with a girl who had just discovered she was pregnant, Music Therapist B recounted a unifying moment that was spontaneously created in the music:

> Then something that just completely took me by surprise, as one of the girls looked toward her, within the same key and just brought in the idea of Jesus Take the Wheel...There was just something so extremely unifying about it.

The participants each remarked upon a specific phenomenon of feeling connected to the music in a heightened way, while working with the adolescent girls in Jamaica. Music Therapist A revealed:

> There is something definitely different about being involved in Jamaican music, in Jamaica. Sharing Jamaican music with a group of people who are living there, for whom it is their culture and their lives. I think just heightened the experience, the motivation, the emotional response, the way that it can have an impact. There is some connection, I think to environmental factors that have a tremendous impact on how we feel about things that we are experiencing.

Music Therapist B additionally shared "The beauty of Jamaica, the rich culture of the music and what it could yield on its own...You just can't be ready to understand their culture till you're in their culture."

Category Four: Impact on the Music Therapist

This concluding category briefly discusses the impact and conclusive insight obtained from the cross-cultural experience. It is summarized in the following three main themes: (a) vicarious trauma; (b) transformative; and (c) widening worldview.

Vicarious trauma. The findings of this study revealed that both Music Therapist A and B expressed powerful feelings and indications of vicarious trauma. Music Therapist A explained, "The emotional response, the way that it can have an impact, that experience was tremendous. And, I think in some ways it is hard to be articulate because it was emotional for so many reasons." Music Therapist B reflected, "We got to experience it all again together, in a very secondary traumatic way."

Transformative. Each of the participants described related moments while working in Jamaica and revealed they were in some way impacted from the overall cross-cultural experience. Music Therapist A indicated that their time in Jamaica was "incredibly touching and rewarding to have been there and to engage in music making with those girls...It's something that I don't want to ever lose." Music Therapist B also referenced specific moments in the cross-cultural experience that he "felt extremely honored to be a part of and I take it with me everywhere I go."

Widening worldview. In the final theme, both participants commented on discovering a new understanding of themselves, their clinical abilities, and remarked on the cross-cultural experience as being a parallel process. For example, Music Therapist A recognized a significant correlation between her own resilience and that of the clients:

> I really had to sort of step outside myself and reshape how I was looking at it and how I was taking it in so I could provide the support that I was there for...New experiences can be really good experiences. And, one of the things we were there to help the girls with was making a change, and being open to change, and being adaptable. I needed to remind myself that I could go through that as well, and be okay, come out okay on the other end.

Similarly, Music Therapist B considered the time working in Jamaica to have been a mutually beneficial and bi-directional experience:

> I think you have to be fairly open, very patient, and these are things that I had to learn about myself...There's just so much more to this culture than I think I know, and that's good, that's a great thing. I am there to learn things too, and they surely taught me a lot about what it means to be a part of music.

Discussion

The purpose of this phenomenological inquiry was to better understand the cross-cultural experiences of music therapists working with adolescent girls with trauma in Jamaica. Individual interviews were conducted with two board-certified music therapists who had previously worked within this setting and population. When asked to describe their cross-cultural experience working in Jamaica with adolescent girls with trauma, the two participants revealed considerably congruent accounts. In particular, four key unifying categories emerged: (a) reactions to the Jamaican context; (b) challenges conducting music therapy; (c) connections through music; and (d) impact on the music therapist. In this section, I use these four categories to shed light on the three major research questions that guide this study.

Research Question One

The overarching research question was: What are the cross-cultural experiences of music therapists working with adolescent girls with trauma in Jamaica? As this is a guiding research question, many of the categories and their respective themes proved informative. For example, category one: reactions to the Jamaican context, was the first to speak to this research question as it concerned how the therapists experienced the overall cultural environment. Under this category, the following themes were presented: (a) culture shock; and (b) countertransference. According to the data, culture shock played a significant role in the music therapists' cross-cultural experience. Witnessing poverty, marginalization, and poor conditions were said to have elicited considerable emotional reactions and a heightened awareness of countertransference in each of the participants. As described in the literature, culture shock is a common reaction when witnessing radically different values and beliefs from one's native culture (Levine & Adelman, 1993). Bigby (2003) asserts that the key element in understanding the dynamics of cultural relations is for the music therapist to have an understanding of their own cultural beliefs. As shown throughout the literature, central considerations such as cultural empathy and cultural perspective, all help toward primary goals when facilitating therapeutic treatment cross-culturally (Pavlicevic, 2004; Stige, 2002; Wheeler & Baker, 2010). Therefore, assessing differing cultural norms and surveying cultural identity, such as values, beliefs, ideals, prejudices and privileges, may help to understand how our personal assumptions are shaped. Furthermore, each participant acknowledged how their perceived feelings of culture related countertransference manifested and in part informed their clinical choices and work.

Another category that helped to address this research question was found in category four: impact on the music therapist. In this category, three significant findings emerged: (a) vicarious trauma; (b) transformative; and (c) widening worldview. These themes all spoke to the effects of working in this cultural context. Specifically, after their time in Jamaica, each participant reported having experienced a sense of vicarious trauma from bearing witness to the living conditions at the safe-house and from hearing the traumatic accounts shared by the girls. Similar to responses of culture shock, witnessing or hearing about others' suffering can vicariously impact the therapist in numerous ways, such as low confidence in professional ability and emotional well-being. Literature in trauma care often discusses the variables of vicarious trauma and therefore, more studies in the field of music therapy are needed towards understanding how to address potential vicarious trauma exposure (Finklestein et al., 2015; Knafo, 2004). Equally, Lowey and Hara (2002) remark on the overall impact on the caregiver and therapist when supporting those who have experienced significant trauma. However, challenges were offset by the meaningful and transformative experiences each music therapist reported having while in Jamaica. As expressed by Music Therapist A, "It was incredibly touching and rewarding to have been able to be there and to engage in music making with those girls...It's something that I don't want to ever lose." Similarly, Music Therapist B shared, "I am extremely honored that I was even a part of that...I take that moment with me everywhere I go, I think about that a lot."

At the end of the interview, each participant reflected upon their overall cross-cultural experience and expressed that it was remarkably impactful. Each participant found his or her time working in Jamaica to be insightful professionally and personally. Similar therapist reactions change can be seen in the literature on trauma care. For example, Gao et al. (2013) describe an extreme array of emotions and attachments when working with client trauma.

Research Question Two

The second guiding research question was: What cross-cultural considerations occur while working with this population? Once again category one: reactions to the Jamaican context provided useful information for this question. For example, respondents expressed that experiencing culture shock was, at times, disruptive to their ability to effectively facilitate therapy sessions. In some cases, participants revealed feelings of uncertainty and questioned their professional abilities, although as Levine and Adelman (1993) and Woodward (2015) suggest, stages of culture shock, loss of self-esteem and setbacks are a common occurrence when working in foreign settings. Noted by Fitzsimons (2016), it is common for music therapists working in different countries to feel "destablished and de-skilled" (p. 85). In addition, each participant expressed the importance of being mindful of how they were reacting to the cultural norms and therefore, allowed their countertransference feelings to inform their clinical work. Music Therapist B shared, "That's a challenging thing for me to just walk in and be like this is a norm and this is what life is." Music Therapist B went onto explain feeling "an overwhelming sense of countertransference that would just jump into me, that I just had to work and understand, as I was getting to know the girls." As illustrated in the literature, such experiences of processing culture have been shown to create differential power dynamics in the therapeutic relationship, leaving the music therapist to question their role (Bolger, 2012; Coombes, 2011; Pavlicevic, 2004; Woodward, 2015). This finding suggests that countertransference when brought into the therapists' awareness, can be an effective tool toward understanding the client's overall world.

In category two: challenges conducting music therapy, the themes addressed spoke directly to the myriad of difficulties that presented and were said to affect the therapeutic process. The four main themes were identified as being: (a) non-therapeutic work environment; (b) language barriers; (c) short-term care; and (d) negotiating cultural stigma. It became apparent that both participants experienced various barriers to their typical way of clinical practice and level of comfort. Each participant remarked that they experienced difficulty facilitating music therapy services within the safe-house. Further, the facility was said to have had numerous factors that impacted the participants' perspectives of a therapeutic space, predominately in relation to the acoustics. Each participant reported additional challenges, such as understanding the common Patois dialect. Within the literature, language appeared to be a prevailing discussion and hurdle within delivering therapeutic services (Lang & McInerney, 2002; Navarro Wagner, 2013, 2015). However, varying music therapy approaches perceive the music experience as a way to access meaningful connection without the need to access verbal speech. According to Navarro Wagner (2015):

> Language is important to create bridges and culturally sensitive settings, but not a barrier for musicking with the participants, because the focus is not so much on interpreting the participant's psyche, but on humans connecting through music (para. 35).

An additional cross-cultural consideration participants identified was the theme of negotiating cultural stigma and ways in which the stigma played a role in how the adolescent girls processed and expressed their trauma. These findings support prior evidence indicating that interactions between an individual and their environment play a role in determining how they may cope with traumatizing experiences (deVries, 1996). Pavlicevic (2002) recommends developing a sense of the client's worldview, including biases and cultural belief systems. Moreover, participants recounted that the trauma impacting the adolescent girls was not simply specific to the Jamaican culture, nor the cultural stigma, but also the subculture of the facility itself. As Music Therapist A noted this concern, "Those teens girls living together under one roof, with such intense emotions, and with a staff that comes in and out. I think an additional separate distinct culture develops there and how different that might look in another country." The cross-cultural literature addresses the idea of subcultures and their own values systems that may emerge within a larger culture (Bigby, 2003). This concept reiterates that the experience of trauma is subjective and encapsulates not only one's identified national or ethnic culture, but rather includes the totality of the individual's life experiences and worldview. However, each participant remarked that many of the exhibited behaviors appeared to be manifestations of typical trauma responses that they had witnessed while working with adolescents in the United States. Interestingly, although both participants acknowledged their cultural identity within the therapeutic context, they seldom remarked upon their race, gender, or age as being a barrier or challenge in music therapy.

Research Question Three

The final guiding research question was: How are those experiences reflected in music experiences? This question was primarily addressed with the third category: connecting through music. Concerning this category, three themes emerged as most relevant for understanding how these experiences translated into the music experiences and included: (a) cultural appreciation; (b) promotion of stability; and (c) unity. For cultural appreciation, participants identified music as being an essential component, embedded within the Jamaican culture. This theme is congruous to sources that remark music as being central to the social and cultural life of Jamaicans (Lewin, 2000). Participants further stressed the importance of preparation prior to the mission trip, and building a culturally sensitive music repertoire. This finding complements literature that expresses a cultural sensitivity approach to music is a significant when working with diverse clientele and provides further knowledge toward cultural norms and/or deviations (Stige, 2002). However, participants equally suggested remaining mindful not to over prepare with preconceived ideas or expectations, thereby allowing the music to unfold organically and authentically in-the-moment. As said by Music Therapist B, "My best preparation was just to be ready for what comes at me and to offer my authentic self musically and interpersonally with them." In addition, participants expressed moments of uncertainty and feeling as if they were within a chaotic environment, as reported by Music Therapist B, "It felt like we were jumping into a little bit of chaos." However, each participant stressed the importance of accessing their "therapeutic sense," being adaptable, and utilizing the qualities of music as a "diffuser" in moments of chaos to best promote a sense of stability. As indicated by Music Therapist A, "The musical choices helped alleviate that chaotic feel at moments."

In addition, study findings suggested that music provided a safe and familiar forum to develop trust and support unity amongst the girls and the participants. Navarro Wagner (2015) conveyed how music can create cultural bridges in a connecting and unifying way. When speaking in regard to the specific music experiences, Music Therapist A remarked that the music was "found to be grounding and calming and foster cohesion." Similarly, Music Therapist B shared that "there was just something so extremely unifying about it." Given the above, one may infer music therapy as being a harmonious and complementary form of psychotherapy in this specific cultural setting. The familiar modality of music, may act as an ideal vehicle to overcome stigma and play a significant role towards health and wellness within the Jamaican community.

Overall, making music with the girls in Jamaica was said to be personally experienced in a "heightened way." Each participant attributed the intense music connection in response to being fully immersed within the culture. Music Therapist A reported, "Sharing Jamaican music with a group of people who are living there...I think just heightened the experience, the motivation, the emotional response, the way that it can have an impact."

Overlapping Themes

While distinct, each of these categories shared conceptual and practical overlap. After deciphering the qualitative data, it became apparent that categorical elements and their respective themes can all be seen in different contexts. For instance, within the data, perceptions of culture shock were present throughout the experience. Yet this theme appeared most prevalent in the category reactions to the Jamaican context. Similarly, it can be seen that feelings of vicarious trauma overlapped throughout the study, although accounts were more frequently reported in regards to the category impact on the music therapist. Furthermore, within the therapeutic relationship between the music therapists and the adolescent girls, various themes emerged regarding difficulties and barriers that are clearly delineated in the category challenges conducting music therapy. However, both participants revealed that connection and cohesion were fostered due to understanding these challenges and cultural differences. In this way, the category connections through the music may be seen as a direct result from the perceived challenges. Interpretations of this overlapping data may reveal a parallel process emerging, whereby the participants' cross-cultural experience mirrors that of the adolescent girls' own processing of their trauma and resiliency.

As a whole, these categories informed the guiding research questions by showing a broad picture of challenge and acclimation, as well as musical connection and increased cross-cultural competency. Each participant reported cross-cultural music therapy and trauma care to be a challenging process of learning and adapting, and ultimately found the experience to be impactful professionally and personally.

Conclusion

The purpose of this phenomenological inquiry was to explore the cross-cultural experiences of music therapists working with adolescent girls with trauma in Jamaica. In this final section, I will summarize the overall study and data findings. Lastly, I will discuss my personal reactions, the limitations of the study and potential implications towards music therapy clinical practice and future research.

Summary

The participants of the study were two board-certified music therapists, who had previously worked within this setting and population. The two participants shared their experiences during an individual 60-minute in-person audio-recorded interview. The collected data was then transcribed into a written document and coded employing thematic analysis (Braun & Clarke, 2006). Significant unifying components emerged through the examination of the transcribed data. These findings were grouped into the presented four main categories: (a) reactions to the Jamaican context; (b) challenges conducting music therapy; (c) connections through music; and (d) impact on the music therapist. Thematic findings and significant material that was discovered through the data analysis was summarized and categorically distinguished.

Overall study results revealed that when working cross-culturally with adolescent girls with trauma in Jamaica, music therapists experience themes of culture shock and vicarious trauma. It appeared however, that these challenges and intense reactions essentially propelled the participants outside of their comfort level and cultural norms. Furthermore, each participant said to have experienced the music in Jamaica in a unifying and heightened way and gained new insight both professionally and personally. In effect, through their cross-cultural experience, participants were able to address preconceived expectations, adapted to unanticipated barriers, and within music located new connections and a widened worldview.

Personal Reflection

Having been myself an active participant on the separate mission trips to Jamaica, I often felt very connected to the participants during the interview process. I was able to relate their cross-cultural experiences to my own. However, at times during the interview, I felt constricted as the participants recalled their individual experiences. This feeling was prevalent when participants described their accounts in detail; I instantly was transformed back to Jamaica, and I could picture the environment and the girls I had worked with. I noticed myself wanting to engage in a verbal dialogue with the participants and was conscious of subduing this urge, not wanting to embed my own ideals and thoughts upon their processing of the experience. As a result, I provided nonverbal gestures and nods to display my attention toward the participants.

Additionally, I found difficulty transcribing and analyzing the data due to my personal connection. I reviewed the audio-recorded interviews several times and transcribed the recordings verbatim. During this immersive process, I often found myself becoming quickly exhausted and experienced significant emotions while writing. Because the aim of my study was to better understand the lived cross-cultural experience, I had asked the participants to be open to reliving intense moments. At that time, I was not fully aware that in doing so, I also would need to revisit those powerful-shared experiences. However, as I continued to investigate each interview closely, I was able to garner new insight and meaning toward my own cross-cultural experience. Overall, I found the process to be beneficial and I discovered a sense of value in contributing through my writing, in an effort to honor the adolescent girls in Jamaica.

Limitations of Study

Although my experience and perspective was not documented within the synthesized data, and protocol was maintained during thematic analysis, personal biases and inherent subjectivity should still be considered. Correspondingly, having worked with each participant separately in Jamaica, the preexisting relationship and shared history may have influenced their retelling of the experience. Conversely, these personal connections might have helped contribute towards the clarification of data and the participants' willingness to share openly. For example, one participant remarked that "if I had been a different interviewer" they might have been conscientious in so readily sharing particular intimate details. Perhaps, the mutually shared experience and trust fostered deeper recollection.

Furthermore, similarities in participant demographics may have not reflected the breadth of perspectives of the diverse music therapy community. For instance, both of the participants were Caucasian music therapists with master's level education, practicing within a humanistic framework and from the same metropolitan area in the United States. Although the participants' gender, age, and years of practice varied, their overall demographic attributes and theoretical orientations were congruent and potentially played a role in their cross-cultural experience. A wider demographic may have provided varied accounts. However, due to the specific setting and population there was a limited sample size of participants. A final limitation was the short-term period of care. Each participant remarked on the experience being too short. Awareness of time may have potentially played a role in how the participants were experiencing the clinical work in Jamaica and might not extend to longer treatment circumstances.

Implications for Music Therapy Clinical Practice

Within the field of music therapy, professional development is continuously stressed. In my opinion, emphasizing and understanding one's own culture, personal meaning, and values, as well as learning about others, is critically important to our growth as informed music therapists. Furthermore, although the AMTA (2018c) standards of clinical practice do adhere to music therapists having a culturally sensitive approach and understanding of multicultural considerations, there still appears to be limited attention toward cross-cultural care. Continued inquiry will help bridge these disparities in practice. Further cross-cultural care and trauma care studies will provide insight towards music therapy education and promote awareness amongst music therapy students, educators and related professionals.

The findings from this study revealed several categories that impact the cross-cultural experience while working with adolescent girls with trauma in Jamaica, which may affect the music therapist both professionally and personally. For example, witnessing the stark socioeconomic disparities in Jamaica proved to be very disruptive to both respondents. From this data, it is possible that more preparation for what to expect will serve to be beneficial. That is, advising music therapists of the challenges of visiting areas characterized by extreme poverty could possibly help them better adjust to potential feelings of culture shock. Moreover, the therapeutic settings offered many challenges that the participants were not prepared for. It could be useful to have opportunities for therapists to learn how to navigate and adjust to such structural factors as poor acoustics or relative social disorder. However, these considerations should be informed by the respective context in which the therapy is to take place. Finally, the data proposed several influential factors that may affect the music therapist while working with client trauma in a foreign country. More explicit support networks that are specifically designed to help therapists could provide much needed guidance from these potential stressors. In addition, future studies may care to investigate the role of supervision and self-care when working within the aforementioned setting and population.

Implications for Future Research

Diverse cultures upon the world stage are beginning to adopt music therapy practice into their model of health and wellness. Current statistics illustrate the expanding typography of music therapy practice across the globe (AMTA, 2018a).

Unfortunately, empirical support is sorely lacking, including basic data such as the number of music therapists working in nonnative countries. Furthermore, there is seldom discussion of cross-cultural considerations while working in a foreign country (Fitzsimons, 2016; Gadberry, 2014; Valentino, 2006, Woodward, 2015). In addition, within the field of music therapy there has been minimal research conducted towards cross-cultural trauma care; current studies that exist tend to focus solely on ethnicity and language, while the role of culture and trauma is much more complex (Wilson, 2013). In addition, there appears to be limited studies exploring the music therapists' experience, with studies predominantly focusing on the client's experience. More research is needed to understand influential factors, such as vicarious trauma and potential stressors that may affect the therapist's ability to provide suitable services.

The findings from this analysis suggest that there is much to learn about cross-cultural music therapy and trauma care from the practitioner's point of view. In light of this, further research is needed to deepen the understanding of the cross-cultural experience while working with client trauma. Further exploration could inform the development of tools and resources to help music therapists navigate the compounding challenges associated with music therapy, cross-cultural care, and trauma care. Some suggestions for future research include replicating this study with: (a) music therapists of different theoretical orientations; (b) an extended clinical period; and (c) extending these findings to include other countries and populations. Concerning supporting therapists in their work directly, additional recommendations for continued studies include: (a) exploring supervision while working with client trauma in foreign countries; (b) methods of self-care while working with client trauma in foreign countries; and (c) gaining the perspective of professionals who have worked in more than one foreign country with client trauma. For the context of Jamaica specifically, more research on overcoming cultural stigma in the region would help inform future practice. Comparison studies may care to examine client attitudes between music therapy and verbal therapy in Jamaica. Lastly, research should look into finding ways to include and train members of the local community as a valuable way of providing cultural insight that could significantly improve the therapeutic process and continuity of care.

To my knowledge, this study is the first of its kind to investigate music therapy practice in the country of Jamaica. It is my hope that the study's findings will entice other professionals and students to continue to exploring music therapy in different cultural contexts. The overarching goal of this study is to serve as a career related educational resource that can enrich current and future therapists toward a wider understanding of cross-cultural music therapy and trauma care both in native and nonnative communities.

References

Ainlay, G. W. (1948). *The place of music in military hospitals.* In D. Schullian & M. Schoen (Eds.), *Music and Medicine* (pp. 1-12). New York, NY: Henry Schuman.

American Music Therapy Association (2018a). *Snapshot of the music therapy profession.* Retrieved from http://www.musictherapy.org/assets/1/7/statprofile11.pdf.

American Music Therapy Association (2018b). *History of music therapy*. Retrieved from https://www.musictherapy.org/about/history/.

American Music Therapy Association (2018c). *American Music Therapy Association standards of clinical practice.* Retrieved from https://www.musictherapy.org/about/standards/.

American Psychiatric Association. (2013). *Diagnostic and statistical manual of mental disorders (5th ed.).* Washington, DC: Author.

American Red Cross (2018). *Sandy response and recovery*. Retrieved from https://www.redcross.org/local/new-jersey/superstorm-sandy-update.

Amir, D. (2004). *Giving trauma a voice: The role of improvisational music therapy in exposing, dealing with and healing a traumatic experience of sexual abuse*. Music Therapy Perspectives, 22(2), 96- 103.

Austin, D. (2001). *In search of self: The use of vocal holding techniques with adults traumatized as children.* Music Therapy Perspectives, 19(1), 22-30.

Bensimon, M., Amir, D., & Wolf, Y. (2008). *Drumming through trauma: Music therapy with post-traumatic soldiers.* The Arts in Psychotherapy, 35(1), 34- 48.

Bigby J. (2003). *Cross-cultural medicine*. Philadelphia, PA: American College of Physicians.

Black, C. (1994). *History of Jamaica*. Kingston, JAM: Carlong Publishers Caribbean Limited.

Bolger, L. (2012). *Music therapy and international development in action and reflection: A case study of a women's music group in rural Bangladesh*. Australian Journal of Music Therapy, 23, 22-39.

Bonny, H. (2002). *Music and consciousness: The evolution of guided imagery and music.* Gilsum, NH: Barcelona Publishers.

Borczon, R. M., Jampel, P., & Langson, G. (2010). *Music therapy with adult survivors of trauma*. In K. Stewart (Ed.), *Music therapy and trauma: Bridging theory and clinical practice* (pp. 39-57). New York, NY: Satchnote Press.

Boyatzis, R. E. (1998). *Transforming qualitative information: Thematic analysis and code development.* Thousand Oaks, CA: SAGE Publications.

Braun V., & Clarke, V. (2006). *Using thematic analysis in psychology*. Qualitative Research in Psychology, 3(2), 77-101.

Bruscia, K. (1998). *Defining music therapy* (2nd ed.). Gilsum, NH: Barcelona Publishers

Bruscia, K. (2002). Preface. In B. Stige (Ed.), *Culture-centered music therapy* (pp. xiii-xviii). Gilsum, NH: Barcelona Publishers.
Child Protection & Family Services Agency (2018). Jamaican child protection & family services agency. Retrieved from http://childprotection.gov.jm.

Coombes, E. (2011) *Project Bethlehem: Training educators and health workers in the therapeutic use of music in the West Bank*. Voices: A World Forum for Music Therapy, 11(1). Retrieved from https://voices.no/index.php/voices/article/view/291/445.

Creswell, J. W. (2014*). Research design: Qualitative, quantitative, and mixed approaches* (4th ed.). Thousand Oaks, CA: SAGE Publications.

Davis, S., & Simon, P. (1992). *Reggae bloodlines: In search of the music and culture of Jamaica.* New York, NY: Da Capo Press.

deVries, M. W. (1996). *Trauma in cultural perspective.* In B. van der Kolk, A. McFarlane & L. Weisaeth (Eds.), *Traumatic stress: The effects of overwhelming experience on mind, body, and society* (pp. 398-413). New York, NY: Guilford Press.

Finklestein, M., Stein, E., Greene, T., Bronstein, I., & Solomon, Z. (2015). *Posttraumatic stress disorder and vicarious trauma in mental health professionals.* Health & Social Work, 40(2), 25–31.

Fitzsimons, B. (2016). *Approaching music therapy in a different county: A literature review on cultural considerations when practicing in a developing country.* British Journal of Music Therapy, 30(2), 83-88.

Gadberry, A. L. (2014). *Cross-cultural perspective: A thematic analysis of a music therapist's experience providing treatment in a foreign county.* Australian Journal of Music Therapy, 25, 66-80.

Gao, T., O'Callaghan, C., Magill, L., Lin, S., Zhang, J., Zhang, J., & Shi, X. (2013). *A music therapy educator and undergraduate students' perceptions of their music for Sichuan earthquake survivors.* Nordic Journal of Music Therapy, 22(2), 107-130.

Jamaica Adolescent Reproductive Health Project (2003). *Fact sheet on youth: Sexual abuse and violence*. Retrieved from http://pdf.usaid.gov/pdf_docs/Pnacy188.pdf.

Juan, T. F. d. (2016). *Music therapy for women survivors of intimate partner violence: An intercultural experience from a feminist perspective*. The Arts in Psychotherapy, 48, 19-27.

Kim, S. (2008). *The supervisee's experience in cross-cultural music therapy supervision.* In S. Hadley (Ed.), *Qualitative inquiries in music therapy* (pp.1-44). Gilsum, NH: Barcelona Publishers.

Knafo, D. (Ed.). (2004). *Living with terror, working with trauma: A clinician's handbook.* Lanham, MD: Jason Aronson.

Kohls, R. (1984). *Survival Kit for Living Overseas*. Yarmouth, ME: Intercultural Press.

Lang, L., & McInerney, U. (2002) *Bosnia: A music therapy service in a post-war environment.* In J. P. Sutton (Ed.), *Music, music therapy and trauma: International Perspectives* (pp. 153-174). London, UK: Jessica Kingsley.

Levine, P. A. (1997). *Waking the tiger: Healing trauma: The innate capacity to transform overwhelming experiences.* Berkeley, CA: North Atlantic Books.

Levine, D., & Adelman, M. (1993). *Beyond language: Cross-cultural communication.* Englewood Cliffs, NJ: Prentice Hall Regents.

Lewin, O. (2000). *Rock it come over: The folk music of Jamaica.* Kingston, JAM: West Indies.

Loewy, J. V., & Hara, A. F. (Eds.). (2002). *Caring for the caregiver: The use of music and music therapy in grief and trauma.* Silver Spring, MD: American Music Therapy Association.

Luckman, J. (2000). *Transcultural communication in health care.* Clifton Park, NY: Cengage.

Mason, P. (2000). *Jamaica in focus*. New York, NY: Interlink Books.

Navarro Wagner, A. (2013*). The experiences of four European music therapists with local disadvantaged children in Gulu's socio-cultural post-war context: A community music therapy point of view.* (Master Thesis.) Aalborg University, Aalborg, Denmark.

Navarro Wagner, A. (2015) *The art of re-framing. Voices: A World Forum for Music Therapy* 15(1). Retrieved from https://voices.no/index.php/voices/article/view/769/676.

Palmer, G. J., Palmer, R. W., & Payne-Borden, J. (2012). *Evolution of counseling in Jamaica: Past, present, and future trends.* Journal of Counseling & Development, 90(1), 97–101.

Pavlicevic, M. (2004). *Learning from Thembalethu: Towards responsive and responsible practice in community music therapy*. In G. Andsell & M. Pavlicevic (Eds.), *Community music therapy* (pp. 35-47). London: Jessica Kingsley Publishers.

Pearlman, L. A., & Saakvitne, K. (1995). *Trauma and the therapist: Countertransference and vicarious traumatization in psychotherapy with incest survivors.* New York, NY: Norton.

Porges, S. W. (2001). *The polyvagal theory: Phylogenetic substrates of a social nervous system*. International Journal of Psychotherapy, 42(2), 123-146.

Ruud, E. (1998). *Music therapy: Improvisation, communication, culture.* Gilsum, NH: Barcelona Publishers.

Samms, K., & Cholewa, B. (2014). *Exploring the context of child sexual abuse in Jamaica: Addressing the deficits.* Journal of Child Sexual Abuse, 23(2), 115-127.

Stewart, K. (Ed.) (2010). *Music therapy and trauma: Bridging theory and clinical practice.* New York, NY: Satchnote Press.

Stige, B. (2002) *Culture-centered music therapy.* Gilsum, NH: Barcelona Publishers.

Stowell, C. R. (2000). *Retention and preservation of African roots in Jamaican folk music*. Retrieved from https://debate.uvm.edu/dreadlibrary/stowell.html.

Sutton, J. P. (Ed.). (2002). *Music, music therapy and trauma: International perspectives.* London: Jessica Kingsley Publishers.

Sutton, J. P., & Backer, J. (2009). *Music, trauma and silence: The state of the art.* The Arts in Psychotherapy, 36(2), 75-83.

Tylor, E. B. (1920). *Primitive culture: Researches Into the Development of Mythology, philosophy, religion, art, and custom.* London, UK: John Murray.

Valentino, R. E. (2006). *Attitudes towards cross-cultural empathy in music therapy.* Music Therapy Perspectives, 24(2), 108-114.

van der Kolk, B. (2014*). The body keeps the score: Brain, mind, and body in the healing of trauma*. New York, NY: Viking Books.

Wheeler, B. (Ed.) (2016). *Music Therapy Research* (3rd ed.). Gilsum, NH: Barcelona Publishers.

Wheeler, B., & Baker, F. (2010). *Influences of music therapists' worldviews on work in different countries.* The Arts in Psychology, 37(3), 215-227.

Wilson, J. P. (2013). *Trauma, culture and metaphor: Pathways of transformation and integration.* New York, NY: Routledge.

Woodward, A. (2015). *Tapestry of tears: An autoethnography of leadership, personal transformation, and music therapy in humanitarian aid in Bosnia Herzegovina*. (Dissertations & Theses. 192). Antioch University, Yellow Springs, OH. Retrieved from https://aura.antioch.edu/etds/192.

World Federation of Music Therapy (2018). *Regional information*. Retrieved from http://www.musictherapyworld.net/WFMT/Regional_Information.html.

The Cyber Orchestra Method's Effect on Attention Span and Self-Esteem in Developmentally Typical 5-6 Year Old Children

Ricardo Hurtado, Samantha Jones, Dr. John Roy Kennedy

Abstract

The purpose of this study was to investigate the effects of the Cyber Orchestra Method as compared to traditional music education activities on attention span and self-esteem of 26 typically developmental children 5-6 years old. The Cyber Orchestra Method is a performance and teaching methodology composed of a combination of conventional music therapy techniques and a unique system of music notation. It includes conventional music theory concepts and piano performance techniques to create orchestra-like acoustics utilizing pre-recorded instrument voices played using electronic piano keyboards. Using a duration recording method, the investigators measured the attention and engagement level of subjects when carrying out traditional music education activities and Cyber Orchestra Method exercises. A pre-test and post-test using the Self-Esteem Child Implicit Association Test (Child IAT) was administered. Results from a two-tailed t-test showed there was a significant difference between the attention span of subjects who used the Cyber Orchestra Method from those who used traditional music education. No significant difference was found on the self-esteem between the two groups.

Literature Review

Self-Esteem

In this review, self-esteem is defined as a person's sense of self-worth or value. Positive self-esteem is associated with positive self-image, feelings of comfort, both in and out of academic settings, and quality musicianship. Low self-esteem is related to negative self-image, emotional issues, and poor musicianship (Culp, 2016).

Music instruction may be positively related to increases in self-esteem (Culp, 2016). Some suggest that music participation in any form, not just music education, may have a positive impact on self-esteem (Darrow, Novak, Swedberg, Horton, & Rice, 2009). Personal satisfaction (including increased self-esteem) was reported as the most important aspect of their secondary band experience and the second most important aspect of their community band experience by adults participating in community band (Cavitt, 2005). Adult piano students also listed the personal benefits of studying piano (including self-esteem) as being significantly important to them (Jutras, 2006).

Costa-Giomi (1999) examined the relationship between music and cognitive abilities by comparing the cognitive abilities between children who received individual piano lessons and children who did not over the course of three years. The children in the study were in the fourth grade, had never participated in formal music instruction, and did not have a piano at home. The results indicated that general and spatial cognitive development were positively affected by music instruction. However, while the group that received instruction had significantly higher scores after one and two years, there was not a significant difference between the two groups after three years. Quantitative and verbal cognitive abilities were not affected by music instruction.

In a continuation of this line of inquiry, Costa-Giomi (2004) examined the same group of students in order to examine the effect of music instruction on academic achievement, school performance, and self-esteem. Self-esteem and academic performance in music classes were positively affected by music instruction, while academic performance in other subject areas was not. The larger increase in self-esteem for students who received musical instruction as compared to students who did not is most likely attributed to the positive attention received from instructors and parents as well as the opportunities to perform for their families and peers. Rickard, Appelman, James, Murphy, Gill, & Bambrick (2012) introduced extra music classes to students in the first and third grades in order to examine the effects of music instruction on social competence and self-esteem. The first-grade group received Kodaly instruction and the grade three group received instrumental instruction in addition to their regular music classes. The music instruction groups for grade one and grade three experienced an increase in self-esteem, while the groups that did not receive music instruction experienced a decrease in self-esteem. This reinforces the results from Costa-Giomi's 2004 study while placing it in the context of the school setting and not just in private lessons.

Roy, Devroop, and Getz (2015) started a concert band program at a disadvantaged public high school in South Africa to compare self-esteem, optimism, positive affect, negative affect, motivation to avoid losses, and motivation to seek gains in students who participated in the band program and students who did not after two and a half years. The study results showed an overall increase in self-esteem observed in both groups. However, there was no difference in self-esteem between students who participated in the band versus students who did not participate in the band observed. This increase in self-esteem for both groups is likely due to the increased age of the participants and not participation in the band.

Mukherjee and Dasgupta (2017) examined the effect of music training on anxiety and self-esteem by comparing adults who had received at least five years of music training and adults who had received no form of music training. Participants who had received musical training had significantly lower levels of anxiety and higher levels of self-esteem compared to those who had not. Male participants were found to have lower anxiety levels and higher self-esteem than female participants. Music can be used as a form of stress relief, which may contribute to lower levels of anxiety. Performance opportunities may boost levels of self-esteem due to the appreciation and admiration expressed by others.

However, this may not be true for individuals with performance anxiety. Anxiety is correlated to low self-esteem. Individuals with low self-esteem may be more likely to experience performance anxiety than individuals with high self-esteem (Ryan, 1998). Negative performance experiences caused by performance anxiety may do more harm than good to the performer's self-esteem.

Certain practices or strategies may affect the effect of music instruction on self-esteem (Darrow et al., 2009). Fostering activities that are at an appropriate level for each student may do this. Activities that are too difficult may discourage or frustrate students (Bissell, 1995; Culp, 2016; Jutras, 2006).

Positive feedback and reinforcement from an instructor can increase confidence and self-esteem. Praising efforts even when mistakes are made can amplify this (Culp, 2016; McAlister, 2012).

Attention Span

Attention difficulties will be defined in this study as issues related to maintaining focus or attention on a task. Attention difficulties may stem from multiple sources, including Attention-Deficit Hyperactivity Disorder (ADHD), learning disabilities, other mental health conditions, nutritional problems, fatigue, and various medical conditions (Williams, 2016). Some hyperactive behavior or attention difficulties are common to all young children, so regardless of cause, teachers should be equipped to cope with attention difficulties in the classroom (McAllister, 2012). In this section, attention difficulties will be discussed in a general sense. Due to its prominence, ADHD will be discussed at length in the next section of this review.

Symptoms of attention difficulties may vary in severity and noticeability (Williams, 2016). Attention difficulties may be related to age. Generally older students can focus for longer than younger students. Older students may also be more adept at filtering out distractions than younger students (Feldman & Contizius, 2016; Millares, 2012).

Attention difficulties pose an issue for the classroom environment, as students who are distracted will likely learn less and potentially distract their fellow classmates (Feldman & Contizius, 2016). Musical activities pose a challenge when working with students with attention difficulties. Playing an instrument requires the student to pay attention to aural, visual, and tactile information simultaneously (Jutras, 2009). This may be a high demand even for students without attention difficulties.

Certain instructional strategies may help assist students having problems maintaining attention in music lessons. Quickly changing activities before attention has time to wane is particularly effective and easy to implement in music lessons (Feldman & Contizius, 2016; Millares, 2012). Self-evaluation and error detection-based objectives are useful for keeping students engaged in more passive activities like listening. Removing possible distractors like electronics can help students stay on track in lessons and while practicing at home (Jutras, 2009).

Transitions between activities are especially prone to losing students' focus. Having students repeat instructions, asking students to listen, and closing the distance between teacher and student are strategies for redirecting their focus. Structuring activities as games or challenges can also serve as a motivation tool for students to stay focused (Millares, 2012).

Music instruction may influence attention span. For students with intellectual disabilities, music may increase attention span (Krebs, 1978). Scott (1992) examined the effects of Suzuki violin lessons, creative movement classes, and preschool activities and classes on attention and preservice behaviors. Eighty children aged three to five participated in the study. Children enrolled in individual Suzuki lessons and group Suzuki lessons scored higher on all attention task variables and on teacher approval than children in creative movement classes or preschool classes and children not enrolled in any type of class. The results of the study indicate that attention behaviors may transfer from instructional settings to other situations.

In order to study the effects of music education versus other leisure activities in elderly people, Seinfeld, Figueroa, Ortiz-Gil, and Sanchez-Vives (2013) compared older adults who had received piano lessons for four months to older adults who participated in other activities such as computer lessons, physical exercise, and others for the same amount of time. The group that received musical instruction had significant improvement in executive function, inhibitory control, and divided attention. Since the control group participated in other activities, these differences can be attributed to music specifically, not just engaging in a stimulating activity.

However, some studies may indicate otherwise. Yeşil and Ünal (2017) examined young adults at a music university in Turkey in order to determine the effects of music training on attention and working memory in adult music trainees. First year students were compared with fourth year students using neuropsychological tests. There was not a significant difference found in attention between the two groups. Data concerning amount of practice time was not collected and could be a contributing factor to levels of attention.

Attention Deficit Hyperactivity Disorder (ADHD)

There are three types of ADHD: inattentive ADHD, hyperactive-impulsive ADHD, and combined ADHD. Inattentive ADHD is marked by difficulties in maintaining concentration, such as not paying attention to detail, ignoring instructions, disorganization, and being easily distracted. Hyperactive-impulsive ADHD is marked by hyper or erratic behavior, such as fidgeting, restlessness, interrupting others, and impatience. Combination ADHD is marked by both difficulty in concentration and hyper behavior (Campbell & Scott-Kassner, 2014; Young, 2009).

Students with ADHD may possess any of the following symptoms: attention difficulties, making careless mistakes, being easily distracted, difficulty maintaining focus or finishing tasks, shifting frequently from one task to another, procrastinating, disorganization, and forgetfulness (Campbell & Scott-Kassner, 2014; McAllister, 2012). ADHD may also be connected to or accompanied by bipolar disorder, conduct disorder, obsessive-compulsive disorder, anxiety or mood disorders, Tourette's disorder, sleep deprivation, or substance abuse (Young, 2009).

Attention difficulties of any degree or origin will have a negative impact on academic performance. However, in addition to its impact on school life, ADHD may impact an individual's home and social life. Socially inappropriate behavior may be a way ADHD impacts a student's classroom experience. Strategies for dealing with socially inappropriate behavior include reviewing the classroom rules prior to each lesson and developing tools for self-reflection. Students with ADHD may not be accepted by their peers due to how their ADHD impacts their social life. If lessons are structured in a way that allows for reduced chances of inappropriate behavior, music lessons can be a great source of positive social interaction and positive affirmation (McAllister, 2012). Music has the potential to positively affect the personal, emotional, social, and work lives of individuals (Acebes-de Pablo & Carabias-Galindo, 2016).

Behavioral strategies and medication are often prescribed by doctors to help with ADHD symptoms (Campbell & Scott-Kassner, 2014; Young, 2009). Music therapy has also emerged as a reliable form of alternative treatment for ADHD. Music therapy exercises that are used in the treatment of children with ADHD include active and receptive listening sessions, moments of interpretation and improvisation, and collaborative work (Acebes-de Pablo & Carabias-Galindo, 2016). While music teachers are rarely trained as certified musical therapists, and therefore unable to plan or implement specific musical intervention, certain instructional strategies may be more effective for students with ADHD (Campbell & Scott-Kassner, 2014). This includes, but is not limited to: maintain a variety of activities, using task-oriented reward activities, allowing students to stand or move around during lessons, providing specific practice strategies in a checklist style for review at home, making an outline of the lesson plan for the student to follow, removing excess distractors from the area, modeling the desired level of energy and attention for the students, providing clear guidelines for the lesson environment, giving immediate feedback for student behavior, asking students to repeat instructions, establishing set routines, sequencing material, encouraging the transfer of previously introduced material to different circumstances, providing frequent feedback, and assessing the student's progress (McAllister, 2012; Young, 2009).

Children with ADHD may be better suited to certain instruments. For example, the string bass (or other instruments that are easy to play while standing) allows students to stand and freely move around, providing an outlet for their excess energy. Large instruments such as the tuba may also be a good fit, as the size makes it harder for distracted students to move around or wander the room (Feldman & Contzius, 2016).

Rather than only being affected by the presence of students with ADHD, music education may also be able to impact the children with ADHD themselves. In a meta-analysis examining the effectiveness of music interventions for children with ADHD, Maloy and Peterson (2015) found that music-based interventions can be moderately effective in improving academic performance of students with ADHD. Since males are more often diagnosed with hyperactive-impulsive ADHD, possible differences in effect between gender should be examined.

The purpose of this study was to investigate the effects of the Cyber Orchestra Method as compared to traditional music education activities on attention span and self-esteem of 26 typically developmental children 5-6 years old.

Method

Participants

There were 26 participants in the study. The participants were divided evenly into the control group and the experimental group. The participants were all five years old except for one, who turned six years old the month before the start of the study. All participants were developmentally typical children with no known impairments or diagnosis that could have hindered subjects' abilities to perform the study's tasks.

Setting and Session Structure

In the experimental group, all participants received one hour of instruction once a week for 18 consecutive weeks. Participants in the experimental group were divided in two groups. Seven participants received instruction on Tuesdays and six subjects received instruction on Wednesdays.

Sessions were held in a standard music educational classroom setting. There were eight electronic piano keyboards in the classroom. The participants were assigned to use one keyboard for every session.

Sessions were conducted by a Cyber Orchestra Method instructor and an assistant. For the first 50 minutes of every session, the instructor and the instructor's assistant would teach the Cyber Orchestra Method to subjects. A series of four video recording time periods followed the first 50 minutes of the session. Each recording time-period was one minute long. The remaining six minutes were used to further instruct subjects on the Cyber Orchestra Method.

In the control group, all participants received instruction at the same time once a day Monday through Friday for four consecutive weeks. Thirteen sessions were held in a standard music educational classroom setting, and three sessions were held in an auditorium-like setting.

Sessions were conducted by a music instructor and two instructor's assistants. Each session was 60 minutes long. For the first 50 minutes, the instructor and the instructor's assistants would conduct general music education activities with subjects. A series of four video recording time periods followed the first 50 minutes of the session. Each recording time-period was one minute long. The remaining six minutes was used to further instruct subjects on the music activity at hand.

For the experimental group, the researcher was the head instructor. A certified music therapist and music educator was the head of the control group. The differences in the meetings of the two groups was due to the participating organization's structural differences between the summer and the fall semesters.

The Cyber Orchestra Method

The experimental group was instructed using the Cyber Orchestra Method. The Cyber Orchestra Method used in the investigation was divided into four chapters and each chapter was divided into sub-sections. The music exercises of the Cyber Orchestra Method used in this study were composed of:

1. A modified music staff framework consisting of the 5th line of a standard music staff and 5 measure-bar lines, which delineated 4 measures total per staff

2. Whole notes, whole rests, half notes, half rests, and quarter notes
3. Numbers on the bottom of the modified music staff, representing the counts for each measure on a 4/4 time signature
4. Capital letters on the top of the modified music staff representing the key-tones for each note notation

Subsection 1.2

Participants were instructed on the Right-Hand Middle C Position. This hand position is comprised of 3 keys: C4, D4, and E4. Using the right hand, participants were asked to place the thumb finger on C4, the index finger on D4, and the middle finger on E4. The ring and index fingers were not used for this hand position.

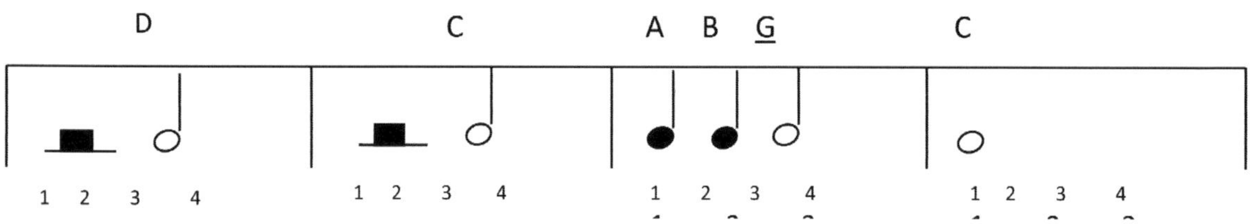

Figure 1.

Section 1

Subsection 1.1

Participants were instructed on how to identify black-keys groups of twos and threes on the keyboard. Each participant was to play all groups of two black keys first, using the index and middle finger of one hand at a time, ascending and descending. Participants were asked do the same with groups of three black-keys using the index, middle and ring finger.

Subsection 1.3

Participants were instructed on how to clap and count whole notes using written exercises. Examples using only whole notes were written down on a classroom board and participants were asked to clap and count them. No electronic pianos were used in this section.

Subsection 1.4

Participants were instructed on how to read, play, and count the Right-Hand Middle C Position music score exercises using whole notes only. Participants were assigned to electronic piano keyboards and provided with written exercises. All participants performed the same exercises.

Section 2

Subsection 2.1

Participants were instructed on the Left-Hand Middle C Position. It is comprised of 3 keys: G3, A4, and B4. Using the left hand, participants were asked to place the thumb finger on B4, the index finger on A4, and the middle finger on G3. The ring and index fingers were not used for this hand position.

Subsection 2.2

Participants were instructed on how to read, play, and count the Left-Hand Middle C Position exercises using whole notes only. Subjects were assigned to electronic piano keyboards and provided with written exercises. All subjects performed the same exercises.

Subsection 2.3

Participants were instructed on how to clap and count half notes. Exercises using only half notes were written down on a classroom board and participants were asked to clap and count the exercises. No electronic pianos were used in this section.

Subsection 2.4

Participants instructed on how to read, play, and count both the Left-Hand Middle C Position and the Right-Hand Middle C Position music score exercises using half notes only. Participants were assigned to electronic piano keyboards and provided with written exercises. All subjects performed the same exercises.

Section 3

Subsection 3.1

Participants were instructed on how to clap and count quarter notes. Exercises using only quarter notes were written down on a classroom board and participants were asked to clap and count the exercises. No electrical pianos were use in this section.

Subsection 3.2

Participants were instructed on how to read, play, and count both the Left-Hand Middle C Position and the Right-Hand Middle C Position music score exercises using quarter notes only. Subjects were assigned to electronic piano keyboards and provided with written exercises. All subjects performed the same exercises.

Subsection 4

Subsection 4.1

Participants were introduced to two new left- and right-hand positions: The *High C* position and the *Low C* position. The Right-Hand High C Position places the thumb finger on C5, the index finger on D5, the middle finger on E5, and the ring finger on F5. At this point a fourth finger position was added to the Right-Hand C Position. The Left-Hand High C Position places the thumb finger on B5, the index finger on A5, and the middle finger on G4, and the Right-Hand Low C Position places the thumb finger on C3, the index finger on D3, the middle finger on E3, and the ring finger on F3. The Left-Hand Low C Position places the thumb finger on B3, the index finger on A3, and the middle finger on G2. Participants were instructed in the electronic piano keyboards in all three C hand positions.

Subsection 4.2

Participants were instructed on how to recognize and count half rests and whole rests.

Subsection 4.3

Participants were instructed on how to read, play, and count music score exercises using a combination of whole notes, whole rests, half notes, half rests, and quarter notes. Participants were assigned to electronic keyboards and divided equally in three groups: The Middle C position group, the High C position group, and the Low C position group. Each group's exercise performed a harmony and melody part to a Cyber Orchestra Method music composition. Headphones were used to isolate the electric piano keyboard sounds. When each group had learned their part, all three groups played their music part simultaneously without headphones while the instructor conducted.

Materials

All electronic keyboards used in this study were 61-Keys. During chapters 1-3 of the Cyber Orchestra Method, pre-recorded string orchestral sounds were assigned to each participant's electric keyboard. During Chapter Four, pre-recorded string, flute, clarinet, and horn orchestral sounds were assigned to different participants' electric keyboards.

For the first eight sessions of the study, the instructor and the instructor's assistant used the first 20 minutes of the session to instruct subjects on how to count notes and rests. This was conducted in a regular classroom setting with no electronic keyboards present.

The next 30 minutes were dedicated to instructing subjects using the electronic keyboards and carrying out the assigned Cyber Orchestra Method exercises. At this time, participants were asked to move into another classroom where eight electronic keyboards were set up.

Measurement Tools

The Implicit Association Test (IAT) used to measure the Self-Esteem results ranges scores from -2 to +2, and are generally interpreted as Cohen's *d* effect size (D). IAT scores of D = 0.2 and below are considered a "slight" IAT effect, IAT scores of D = 0.5 and above represents a "moderate" IAT effect size, and the IAT scores of D = 0.8 and above represent a "strong" IAT effect. The same convention holds true for the effects obtained in the negative direction.

In addition, an Interval behavioral recording method and a Duration behavioral recording method were used to measure the length of time subjects concentrate to attend a specific giving task. As part of this research, design and guidelines to procure data accuracy and resource efficiency, neither the instructors from both the experimental group and the control group were allowed to intervene in favor to promote attention engagement in subjects during recording sessions.

Control Group Instruction

The control group's instruction represented a scope of age-appropriate general music education activities that involved singing, instrumental playing, moving to music, music listening, music theory, and learning different non-musical skills through music. This music activities were conducted during the first 50 minutes of the session. During the recording periods (last 10 minutes of the session), subjects were asked to carry out specific exercises. The exercises consisted of note identification and counting and clapping to specific rhythms consisting of whole notes, half notes and quarter notes. For these exercises, the same modified staff used with the experimental group was used (illustrated in Figure 2.), with the exception that key-tone letter names were excluded.

Figure 2.

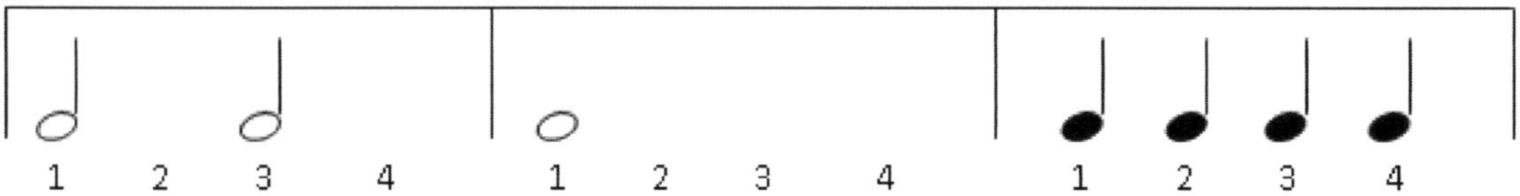

Music education exercises were presented to subjects either on a projector screen or on printed material. Participants were instructed to carry out each exercise before and after each recording period.

Participants were to count each note by pronouncing the syllable "Ta", while holding the "a" sound for the duration of each note. Participants were instructed to clap each time they said "Ta". An example of the exercises used in the control group is represented in Figure 3.

Each exercise was timed and recorded for one minute. The instructor dedicated time before recording to provide instruction to participants. The same procedure was used for all exercises.

Figure 3.

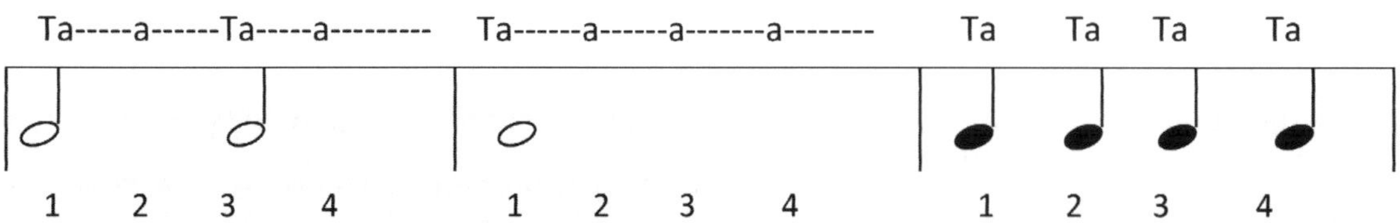

Results

A two-tailed t-test was used to determine if there was any statistical significance between the results from the experimental group and the control group in the participants' attention span and self-esteem. There was a significant difference between the attention span of the experimental group (M = 95.56, SD = 8.18) and the attention span of the control group (M = 37.90, SD = 21.25), T = 9.13, P < .025. The lowest score for the experimental group was 69.11% (n = 9), and the highest was 99.82% (n=4). Control group scores ranged from a low of 5.52% (n=5) to a high of 89% (n=12). Using a confidence level of 0.05 (5%), the results of the data analysis indicate that the probabilities of getting the same results due by chance is less than 0.025. The Null Hypothesis was therefore rejected. The experimental group showed a significantly higher level of attention span than that of the control group.

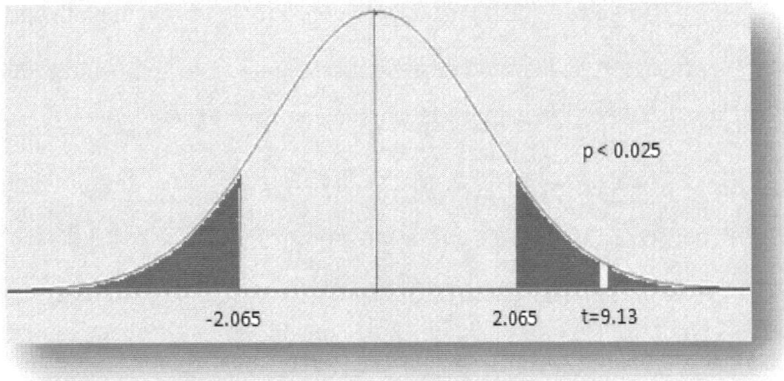

Figure 4. Attention Span T-Test

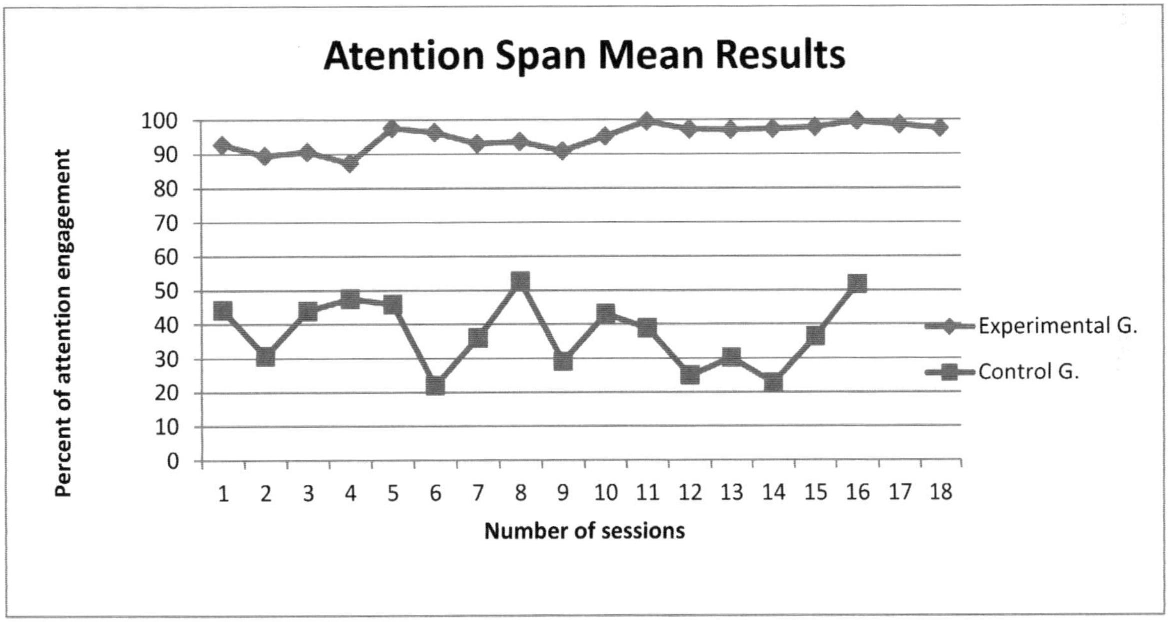

Figure 5. Attention Span Means

The experimental group measurement scale establishes an overall increase of 5% in the attention span of subjects throughout the entire study. The control group had much more variability from session to session without an identifiable trend. From the first to the fourth session, there is a slight decrease from a 93% to 87% attention span. In the fifth session, there is a significant increase from 87% to 98% attention span, and which is followed by a slight decrease to a 91% up until the 9th session. From the 10th session, the attention span is increased up to 100% and maintained with a mean of 98% up until the 18th session. The experimental group displays a higher and more consistent attention span proficiency while the Control Group exhibits a lower and a more irregular attention span measurement outcome.

The lowest score for the experimental group was -0.97 (n = 4), and the highest was 0.75 (n=9). Control Group scores ranged from a low of -1.47 (n=7) to a high of 0.92 (n=8). Using a Confidence Level of 0.05 (5%), the results of the data analysis indicate that the probabilities of getting the same results due by chance is more than 0.025. The Null Hypothesis is therefore not rejected.

Figure 6. Self-Esteem T-Test Data Results

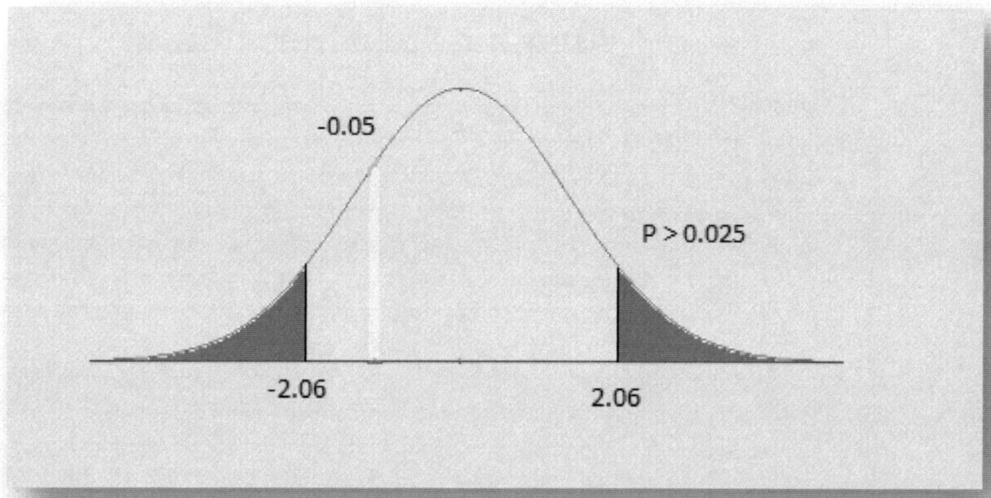

Discussion

There was a statistically significance increase of attention span in children receiving the Cyber Orchestra Method as compared to children receiving only conventional music education instruction. However, there was not a statistically significant difference between children receiving the Cyber Orchestra Method and children receiving only conventional music education instruction for self-esteem.

Some considerations must be brought about when examining all factors that might have influenced the end results of the study. The difference between layout structures of sessions between the experimental group and the control group. While participants in the experimental group received a single session once a week for 18 weeks, participants in the control group received 16 sessions all within one month.

The difference in time-period from one session to the next, and the difference of the overall time of the study between the two groups may have had an effect when contrasting participants' memory retention rates and performance of music exercises. The control group and experimental group also had different instructors and a different number of assistants. It is recommended that future research replicate this study with identical conditions between the experimental and control groups.

Another factor that could have impacted the results is the difference in instructional methods between the two groups. While participants in the experimental group used electronic piano keyboards to execute the music exercises provided, participants in the control group carried out their exercises by clapping and counting to music notation rhythms. It is possible that participants in the experimental group would be more engaged and attentive due to the presence of the electronic keyboards in contrast to commonly used procedures in a typical music education setting. Further development of the method, and other similar methodologies should be done in order to fully assess the benefits that music-based intervention of this kind can bring on the developmental child is recommended.

The limited instructor's proximity to the participants and the limited instructor's ability to change activities or to implement behavior strategies due to research videotaping protocol in this study may also be taken into consideration when assessing results. However, this strategy was implemented to procure precise report findings and objectivity, in order to maintain commonly accepted codes and norms for research integrity.

References

Acebes-de Pablo, A., & Carabias-Galindo, D. (2016). El alumnado de primaria con Trastorno por Déficit de Atención/Hiperactividad (TDAH): la Musicoterapia comoherramienta integradora dentro del contexto del aula de música. *Revista Electrónica De LEEME, 38,* 1-16.

Bissell, P. M. (1995). Keyboards launch students into music. *Music Educators Journal, 82*(1), 29-31.

Campbell, P. S. & Scott-Kassner, C. (2014). *Music for exceptional children*. In Music in Childhood:From preschool through the elementary grades (4th ed.) (pp.404–430). Boston, MA: Schirmer Cengage Learning.

Cavitt, M. E. (2005). Factors influencing participation in community bands. *Journal of Band Research, 41*(1), 42-59.

Costa-Giomi, E. (1999). The effects of three years of piano instruction on children's cognitive development. *Journal of Research in Music Education, 47*(3), 198-212.

Costa-Giomi, E. (2004). Effects of three years of piano instruction on children's academic achievement, school performance and self-esteem. *Psychology ofMusic, 32*(2), 139-152.

Culp, M. E. (2016). Improving self-esteem in general music. *General Music Today, 29*(3), 19-24. doi:10.1177/1048371315619962

Darrow, A., Novak, J., Swedberg, O., Horton, M., & Rice, B. (2009). The effect of participation in a music mentorship program on the self-esteem and attitudes of at-risk students. *Australian Journal of Music Education, (2*), 5-16.

Feldman, E. & Contizius, A. (2016). Inclusion - instrumental music for students with disabilities. *In Instrumental Music Education: Teaching with the musical and practical in harmony* (2nd ed.) (pp. 233–238). New York, NY: Routledge.

Jutras, P. J. (2006). The benefits of adult piano study as self-reported by selected adult piano students. *Journal of Research in Music Education, 54*(2), 97.

Jutras, P. (2009). May I have your attention, please.... *Clavier Companion, 1*(2), 4-5.

Krebs, B. L. (1978). Music education for the mentally handicapped: A summary of selected related literature with conclusions and recommendations. *Bulletin of the Council for Research in Music Education, (57)*, 11-22.

Maloy, M., & Peterson, R. (2014). A meta-analysis of the effectiveness of music interventions for children and adolescents with attention-deficit/hyperactivity disorder. *Psychomusicology: Music, Mind, And Brain, 24*(4), 328-339.

McAllister, L. S. (2012). Positive teaching: Strategies for optimal learning with ADHD and hyperactive students. *American Music Teacher, 61*(4), 18-22.

Millares, M. D. (2012). Attention span in the 5- to 7-year-old music student. *American Music Teacher, 61*(5), 20-24.

Mukherjee, A., & Dasgupta, S. (2017). Role of music training in anxiety and self-esteem of adults. *Indian Journal of Community Psychology, 13*(2), 207-214.

Rickard, N. S., Appelman, P., James, R., Murphy, F., Gill, A., & Bambrick, C. (2013).Orchestrating life skills: The effect of increased school-based music classes on children's social competence and self-esteem. *International Journal of Music Education, 31*(3), 292-309.

Roy, M. M., Devroop, K., & Getz, L. (2015). Improvement in South African students'outlook due to music involvement. *Music Education Research, 17*(4), 465-479.

Ryan, C. (1998). Exploring musical performance anxiety in children. *Medical Problems of Performing Artists, 13*(3), 83-88.

Seinfeld, S., Figueroa, H., Ortiz-Gil, & J Sanchez-Vives, M. V. (2013). Effects of musiclearning and piano practice on cognitive function, mood and quality of life in older adults. *Frontiers in Psychology, 4*, 1-13.

Scott, L. (1992). Attention and perseverance behaviors of preschool children enrolledin Suzuki violin lessons and other activities. 40(3), 225-35.

Williams, A. D. (2016). The development of a music program to improve the attention span of school-aged children (Doctoral dissertation). Retrieved from ProQuest Information & Learning. Capella University, United States.

Yeşil, B., & Ünal, S. (2017). An investigation on the effects of music training on attention and working memory in adults. *Anadolu Psikiyatri Dergisi, 18*(6), 531-53.

Young, M. (2009). Creating successful learning environments. *Clavier Companion, 1*(2), 52-55.

The Effects of Music Therapy on Second Language Acquistion When Used as a Supplemental Tool to Traditional English-as-a Second Language Instruction of Adult Immigrants in a Community College

Aisling Mohini Manison, John Roy Kennedy,

Lauren Viljama, & Kristopher Hong

Abstract

Music has long been recognized as being closely related to language and very effective for teaching. ESL teachers are always exploring creative methods for increasing second language acquisition in their students. While music is frequently used for children at the elementary levels, it is less commonly used for adults, who struggle the most with language learning. The purpose of this study was to investigate the effectiveness of supplemental music therapy activities for second language acquisition (SLA) compared to the effectiveness of traditional classroom methods alone. Subjects were comprised of 38 adult ESL students enrolled in a community college in the Southeastern part of the United States. Students were divided into an experimental group (n = 23) and a control group (n = 15). The control group received traditional ESL instruction only, whereas the experimental group received both traditional ESL instruction and supplemental ESL music therapy instruction in the last 30 minutes of their normal class time twice per week. As a standardized assessment, the BEST Literacy Assessment was used to compare the gain scores of the experimental group to the control group, as well as to analyze the improvement of the experimental group's test scores from pretest to post-test. In addition, Hurley and Tinajero's Speaking Checklist and Story Retelling Inventory were used as a formative assessment to assess the expressive English language skills of students in the experimental group at the beginning and end of the duration of the supplemental music therapy instruction.

Introduction

Music is a time-tested tool for learning, even before the written word. Tribes in Africa and clans in India have passed down stories through music for generations. Before babies can even speak, they communicate through sound in the form of babbling (Levitin, 2006; Honing, 2011). Thus, music is innately linked to communication, making it an essential tool for language learning and acquisition. Many elementary school English language teachers have recognized this fact, as demonstrated by the abundance of resources on social media for creative activities involving songs and movements for learning. In fact, many of the language teaching methodologies that have been developed over the past century are very similar to activities used in music therapy. However, the use of music for language learning has remained localized for use with young children, while adults continue to struggle with traditional language learning methods. The following literature review explores the use of music therapy techniques for helping adults to learn English more efficiently. Firstly, there will be a summary of important ideas and principles in teaching a foreign language and how music could be helpful. Secondly, the application of music therapy techniques will be discussed.

Firstly, consider some key principles of language acquisition and how they relate to music. Automaticity is one of the most important characteristics of acquired language, as this means that the language that the infant hears has been saved to the subconscious. The complexity of language makes conscious management of language at a fluent level very difficult, if not impossible. "Efficient second language learning involves a timely movement of the control of a few language forms into the automatic processing of a relatively unlimited number of language forms. Overanalyzing language, thinking too much about its forms, and consciously lingering on rules of language all tend to impede this graduation to automaticity" (Brown, 1994, p. 17) Steven Krashen proposes that the unconscious, automatic approach is what differentiates "acquisition" from "learning" (Brown & Palmer, 1987, p. 6-7). The acquisition of music is well-known for automaticity and unconscious acquisition, "Man absorbs sounds often unconsciously, and retains music in his mind, perhaps even very short passages which can recall memories of facts, moods or feelings associated with it." (Alvin, 1975, p. 81). In order to sing or play long, complex songs, musicians "overlearn" their repertoire so they can rely mostly on "muscle memory." Secondly, meaningful learning is the linking of new information to existing knowledge and memories, unlike rote learning, which means learning small, scattered bits of information that have little connection to existing schema. Meaningful learning will more likely result in long-term retention (Brown, 1994). However, creating a meaningful learning environment can be difficult. As highlighted by Alvin (1975) above, music has the ability to create a meaningful environment and to link new information to a long-term memory (p. 81). Another principle is the anticipation of reward (Brown, 1994, p. 19). Students' learning is enhanced when a goal is set before them. Music offers bountiful "opportunities for socially acceptable reward and non-reward" (Gaston, 1968, p. 33).

In addition to cognitive principles, language acquisition has many affective principles. Firstly, the development of a "language ego" makes language learners very sensitive, fragile, and needy of what Brown (1994) calls "tender loving care" (p.22). Music creates a highly supportive environment that can be used to cushion, protect, and support students (Gaston, 1968, p. 33). Another principle is that of boosting and maintaining students' self-confidence, which again indicates that music fulfills by providing "successful experiences," "feeling needed by others," and "enhancement of esteem by others" (Brown, 1994, p.23; Gaston, 1968, p. 33). The third affective principle of risk-taking has substantial weight in the theory of language acquisition (Brown, 1994). According to the cognitive theory, students learn from receiving input, processing it, producing output, and receiving positive or negative feedback that either confirms or corrects their understanding (M. Volynsky, personal communication, 2015). An important step in this process, however, is producing output, which will not occur if students are afraid to take risks. Participating in music-making is a risk in and of itself. Thus, music creates a relaxing environment in which students become accustomed to taking risks when producing language. Lastly, students become familiar with the culture as well as the language, since the two are so closely connected (Brown, 1994; Gaston, 1968). Music is inherently part of culture and the target language (TL).

Lessons Learned from Experiments in Language Acquisition

Throughout the past century, linguists have theorized and tested several different methodologies of teaching for language acquisition, each with its successes and failings. Firstly, the audiolingual method (ALM) was developed during WWII out of a need for fast fluency for the U.S. military (Brown, 1994). As a rote method, the goal was to "overlearn" (p. 58) for habit-formation. First, students heard a full conversation role-played by the teacher. Then, they repeated and memorized the dialogue through pattern drills. They were not allowed to write anything until they concluded the drills. Good pronunciation and error-free language was emphasized. No grammar explanation or translation was provided, so students did not fully understand the semantics or syntax they were saying. Students, however, were able to begin speaking very quickly. The use of music with lyrics has the potential to include the repetitiveness that made ALM effective.

Later, Community Language Learning (CLL) came to birth (Brown, 1994). The creator, Charles Curran felt the need to create a nurturing and supportive environment focused on the learners. Thus, the format of the class was similar to that of a music therapy session, with the class sitting in a circle and the teacher outside the circle. Beginners spoke in the L1 and the counselor translated into English for the whole group. Then the "clients" (p. 59) repeated back the translated sentence. Eventually, the students began speaking more and more phrases and sentences directly in English. Clients often helped each other with communication, which added to the supportive atmosphere. Overall, CLL eliminated many of the threats that impede SLA: the "all-knowing teacher" (p. 60), the fear of making mistakes in front of the class, competition, alienation, and inadequacy. The supportive group atmosphere is a natural effect of music-making in group musical activities and is known to maximize the potency of music (Gaston, 1968, p. 27).

Another method developed by Georgi Luzanov (1978) called "Suggestopedia" capitalized on the importance of relaxation and giving control over to the teacher. Music played a vital role in relaxing clients, especially Baroque music, which Luzanov believed induced increased retention "due to an increase in alpha brain waves and a decrease in blood pressure and pulse rate" (p. 61). For example, Luzanov described a concert session which opened with quiet listening to music until the teacher began to read text with voice modulation synchronized to the music as students followed along. Then, there were a few minutes of silence. The music began again, and the teacher read again, except this time students just listened. At the conclusion, students left silently and were given no homework except to read before bed and in the morning. Disadvantages of this method were the questionable experimental data, practicality of resources, and the focus on purely memorization instead of understanding. As Brown (1994) says, however, we learned "that deliberately induced states of relaxation may be beneficial in the classroom." (p. 62). Paired with various other effective language acquisition techniques, music's aesthetic qualities can be used for making students comfortable in the learning environment. In fact, in a research study on music therapy as an intervention for middle school ESL skills, guitar strumming was used as a background stimulus and for filling the silence as students worked on an activity (Kennedy, 2005).

Like Suggestopedia, Gattegno's Silent Way explored learning in unique ways (Brown, 1994). Called "discovery learning" (p. 63), it relied purely on the use of inductive processes. The goal was for learners to develop independence, autonomy, and responsibility. Students figured out concepts together, while the teacher acted as a stimulator. This meant single word stimuli and short phrases and sentences provided once or twice and limited corrective feedback. Beyond this, students were on their own to infer and learn the meaning of words and phrases. This resulted in distant relationship and a lack of understanding between the teacher and students and impeded the communicative atmosphere. However, the emphasis on discovery learning is an important aspect that may be facilitated by music, especially with its flexibility for self-expression (Gaston, 1968).

In the 1960s, James Asher began developing a methodology, Total Physical Response (TPR), based on the "trace theory" (p. 64) which claimed that motor activity stimulated memory and the listening approach that children use to learn their L1 (Brown, 1994). Children listen and respond to their environment through physical means, also known as "listening accompanied by physical response" (Brown, 1994, p. 64). Similarly, students who learn through TPR listened, acted, or made gestures in response to commands or questions given by the teacher. However, this technique was limited to learning concrete operations and was too difficult to use for abstract expression of ideas. Nonetheless, it shows the benefits of using movement with language learning activities. For adults, this idea is significant. In a study, Asher concluded, "When adults learn a second language under the same conditions as children [that is, listening and moving], the adults are superior ... [in] listening fluency" (Krashen, 1982, p. 83). Interestingly, in another study conducted by Asher, it was found that following movements with a model to follow in training resulted in lower listening comprehension scores than individually listening to commands and physically acting during retention tests (Krashen, 1982). Thus, while movement is effective, it is best when the movement is initiated by the learner themselves, usually through command. Music generates physical response in the form of reflexive movements and energy (Alvin, 1975). In addition to this, song lyrics are also often full of commands.

Child-Adult Differences in Language Acquisition

There is a big debate on child-adult differences in language acquisition. Studies show that adults excel in short-term language acquisition in comparison to children, but children outperform adults in long-term language acquisition. The following are some ideas for the potential reasons for the differences.

Firstly, it is a popular belief that the language that children learn is much simpler than the language that adults need to learn (Krashen, 1982). "Speech address to children is highly redundant, syntactically simple, repetitious, related to objects or events in the immediate environment, and parceled out in short utterances" (Krashen, 1982, p. 47).

An issue with this reasoning, however, is that children learn from adults, not other children, and thus receive the same input as adults (Krashen, 1982). In any case, music is adaptable for various levels, from lyrics with very simple language to very complex, authentic language.

Some authors say that the motivation is also different. Children have an intense desire to fit in with their peers (Krashen, 1982). Being in a group with their peers gives adults the opportunity to compare themselves with others in order to assess their expressive language level. According to Sears, "If appropriately provided, [this comparison] may stimulate the individual to further accomplishment" (Gaston, 1968, p. 41). Thus, the effect of the group can be very powerful.

Still others say that repetition and mimicry is easier for children because of better memory (Krashen, 1982). There is no doubt that memory plays an essential role in language acquisition. Moore (2008) talks about two different types of memory: declarative, which is the conscious learning to which Krashen referred, and procedural, which is the automatic, subconscious acquisition that is the ultimate goal for language. In her study, however, Moore focuses on the use of declarative memory for learning. If Krashen's theory of unconscious acquisition holds true, then focus should be on procedural memory, so that language becomes a habit, not just vocabulary lists and grammar rules. Once students have ingrained the habit, then they can delve into analysis of the language. This process resembles the educational model that native speakers undergo when earning their first language as an infant through unconscious acquisition and conscious study in school. Several music therapy techniques are useful for facilitating the acquisition of language into procedural memory. These include Musical Mnemonics Training (MMT), Musical Speech Stimulation Therapy (MUSTIM), and Rhythmic Speech Cuing (RSC) (Davis, Gfeller, & Thaut 2008).

Furthermore, Krashen (1983) spoke of what he called "Din in the Head" (p. 41); that is, the involuntary rehearsal of language, which is an activity that ESL students at all levels utilize (Schunk, 1999). Songs also have a tendency to get "stuck in the head" or become "brainworms" (Sacks, 2007, p. 45). Sacks (2007) explains that this is because "music has entered and subverted a part of the brain, forcing it to fire repetitively and autonomously" (p. 44) Murphey suggests harnessing this characteristic of music to facilitate the Din phenomenon, "Song may act as a LAD [Language Acquisition Device] activator, or be a strategy of the LAD in the ontogenetic development of language" (Murphey, 1990, p. 53). "The L2 [second language] research question is whether music and song can trick the LAD into a Din mode that would process more communicative speech" (Murphey, 1992, p. 773). Thus, setting ideas to music can be very effective for memory. However, in a study conducted by Gfeller (1983), students informally commented that familiar tunes are recalled more easily than unfamiliar ones. Thus, song choice is important and might involve the use of songs that are native to the learner's country (Moore, 2008).

Pronunciation and overcoming strong accents is often a great difficulty for adult learners. Lenneberg notes that "foreign accents cannot be overcome easily after puberty" (Krashen, 1982, p. 205). Pronunciation of specific phonemes can be targeted in singing activities, as was done in one study (Kennedy 2008, p. 96). Also, Vocal Intonation Therapy (VIT) is useful for improving pronunciation (Davis, Gfeller, & Thaut 2008, p. 269). So far used with people in neurological rehabilitation, VIT addresses inflection, pitch, breath control, timbre, and dynamics which simulate characteristics of speech, including prosody, inflection, and pacing.

As mentioned before, adults and children are very different in the amount of physical movement they use on a regular basis. "Children may learn the new language in play situations when utterances are synchronized with physical movement. It may also be that adults learn the new language in static, non-play situations in which their kinesthetic system is not active and not synchronized with speech transmission or reception" (Krashen, 1982, p. 4**)**. In one study, it was found that using singing paired with gestures enhanced vocabulary retention (Schunk, 1999). This demonstrates the effectiveness of adding movement and music to language acquisition.

This idea that adults need a formal environment, not a play environment, adversely affects learning in other ways as well (Krashen 1982). Chambers (1999) cites some of the factors that lead to "reluctant learners," one of which is that the "course content and teaching methods are insufficiently stimulating or relevant to pupils' needs" (p. 7). If lessons are not engaging, learners will not engage in language learning and have a negative attitude towards learning (Murray & Christian, 2011). In a survey of secondary school students in the U.K., students felt that activities like "watching television or video program in French," "reading French magazines or storybooks by yourself," "writing a made-up story in French," and most significantly, "singing French songs" were not part of the formal lessons (Chambers, 1999, p. 122). Some of these activities, however, were among the "most liked" learning activities (Chambers, 1999). Music can break down the affective barriers of a formal environment and make it fun and engaging for students with nontraditional activities (Gaston, 1968). The key to this is interesting input and requiring plenty of output from students.

The power of music to break down affective barriers was seen in the attitude change in the middle school students mentioned before. "During the first half of the semester the 7th grade group often sat in the activity circle with crossed arms, displayed negative facial expressions and body language or turned away from the circle during music therapy activities." (Kennedy, 2005, p. 256). However, after a few sessions, student participation and on-task behavior gradually increased. Thus, "over the course of treatment engagement of the 7th grade students in active learning through music therapy activities were effective in piquing their interests in learning English as a Second Language" (Kennedy, 2005, p. 256).

Perhaps the most cited reason is brain plasticity. It has been said that the brain is more plastic when younger.

Chomsky's Universal grammar theory states that we are all born with the propensity to learn any language, no matter where we are born, as long as we are immersed (Murray & Christian, 2011). However, when we get older, our brain locks into a certain language and way of thinking. Learning a foreign language while young is replaced by other habits. By adolescence, these habits are too deeply ingrained and create enough interference that mastery of another language becomes difficult, if not impossible" (Krashen, 1982).

Another possibility is that the right-hemisphere of the brain no longer involved for language. "It seems, then, that until a certain point, the right hemisphere, not normally differentiated for language, has the capacity to fully develop the linguistic potential generally left unrealized" (Krashen, 1982, p. 32). However, "While the pattern of *occurrence* of aphasia gives evidence of lateralization, the pattern of *recovery* from aphasia can give information on language learning capacity" (Krashen, 1982, p. 32-33). Congresswoman Gabby Giffords, who acquired aphasia (damage on the left side of the brain due to a gunshot wound), recovered speech by transferring this task to the right-brain, made possible through music therapy techniques (Mossie, 2011). This leads us to wonder, if music may be used to give adults that same flexibility that children have of using the right side of the brain for language (Krashen, 1982).

Some linguists say that adults' ability to think abstractly about language and its construction puts them in a superior position to children (Krashen, 1982). Thus, adults are often expected to learn language much faster than children. Interestingly, this ability to think in terms of grammar leads to what Krashen (1982) calls the "L1 plus Monitor Mode," in which "performers simply utilize the surface structure of their first language, and then employ the conscious grammar as a Monitor to make alterations to bring the L1 surface structure into conformity with their idea of the surface structure of the second language" (p. 210). This produces very fast results at early levels, much faster than children at that level. However, as the grammar becomes more and more complex, the conscious approach becomes more difficult and slower (Krashen, 1982, p. 211). As Krashen (1982) summarizes, this means "formal operations relate directly to the ability to *learn* language consciously" (p. 209). If the goal is to learn language subconsciously, music can be very useful.

The greatest enemy of adults for SLA could well be self-consciousness (Krashen, 1982). "Adults are more likely than children to be concerned about minor imperfections in lexical choices, syntax, or articulation either in themselves or in others" (Krashen, 1982, 46-47). Krashen (1982) hypothesizes that "at around puberty, the affective filter is strengthened" (216). This strengthened affective filter is evident in the comment of a 13 year-old student in the U.K., "I feel scared to answer. If I make a mess, people will laugh at me" (Chambers, 1999, p. 125).

Self-consciousness often holds adults back from producing output, which is a tenet of the cognitive theory. If learners do not produce output, they will not receive positive or negative feedback that enables them to perfect their understanding. Krashen (1982) links the formal operations to this heightened affective filter (p. 217). As seen during puberty, the ability to think abstractly enables adolescents to think about what others think of them, which often becomes an obsession, a feeling that they are constantly being judged (Krashen, 1982). Thus, he concludes "formal operations are no friend of adult second language acquisition" (Krashen, 1982, p. 217). There are several reasons why music and the music therapy environment can be effective for overcoming this obstacle. Firstly, students are in a supportive environment of people who will not laugh or mock. Rather, music implicitly sends a message of peace. It is nonthreatening. Secondly, there are many more opportunities for verbalization without being in the spotlight, such as in group singing and chanting. Lastly, making music inherently requires overcoming self-consciousness. Once students can make music, they may find speaking to be very easy in comparison (Gaston, 1968).

Commonalities in Music and Language Brain Function

Research has shown that speech and music share mechanisms for sound discrimination. For both speech and music, no sound is precise. Thus, listeners need "to develop a mental framework that allows them to extract a small number of meaningful categories from acoustically variable signals" (Patel, 2008, p. 78). If this is the case, states Patel, then improving the ability to discriminate sound categories in one domain should improve the same ability in the other domain. In one study conducted with 50 Japanese adult learners of English living in the U.S., participants were given both language and musical tests.

The results clearly revealed that musical ability was predictive of "L2 receptive and productive phonology, in other words, to that aspect of language most directly related to sound categorization skills in perception" (Patel, 2008, p. 79). Natural speech involves pitch, and it can be difficult or impossible to understand meaning without intonation (Hodges & Sebald, 2011). Accented syllables give meaning to words and inflection in phrases and sentences facilitate understanding overall meaning

It is a well-known fact that vocal music is heavily influenced by the rhythm of speech (Patel, 2008). Interestingly, rhythms in instrumental music can also be heavily influenced by the rhythms in the L1 of the composer. In a study comparing the rhythm of English and French speech to rhythms in English and French music, it was found that the English and French speech were very similar to the English and French music, respectively (Patel, 2008). Thus, this might lead us to hypothesize that regularly listening to vocal or instrumental music composed by a native-speaker of the target language could help establish a subconscious familiarity with the rhythm of speech in the TL. Furthermore, Patel (2008) finds that there is evidence that speech and music involve the same type of brain processing for recognizing rhythmic phrases. Another source states, "Rhythmic activities in the acquisition of language are so important that they form the basis for acquiring cognitive expectancies and for interrelating cognition and effect" (Hodges and Sebald, 2011, p. 151).

Patel (2008) proposes the direct-route hypothesis which is "the idea that implicit learning of prosodic patterns in one domain (ordinary speech) influences the creation of rhythmic and tonal patters in another domain (instrumental art music)" (p. 224). This would explain why instrumental music that is not derived or inspired by folk music still often reflects the melodic patterns of speech. "This research suggests that that although musical and linguistic syntax have distinct and domain-specific syntactic representations, there is overlap in the neural resources that serve to activate and integrate these representations during syntactic processing (the "shared syntactic integration resource hypothesis" [SSIRH])" (p. 297).

When meaning is viewed in the light of a direct correlation of one expression to a very specific, non-subjective object or event, then music has very little meaning in comparison to speech. If meaning, however, is understood as the "perception of an object/event [bringing] something to mind other than the object/event itself," then music becomes very meaningful (p. 350). Thus, the gap between meaning in speech and music shrinks. Patel encourages comparative research on music and language.

Experiments in Music Therapy for ESL

Thus far, there have only been a few applications of music therapy for language learning. In 1999, Schunk conducted a study on the use of singing paired with signing for improving receptive vocabulary skills (Schunk, 1999). The study involved four groups who learned in four different ways: sung text paired with signs, spoken text paired with signs, sung text, and a control group of spoken text only (Schunk, 1999, p. 110). In her results, she found that the groups that used sung text paired with signs and spoken text paired with signs did substantially better with vocabulary recognition than the other two groups. Furthermore, "children who actively engaged in singing or speaking new vocabulary words paired with signs performed significantly better than did children who rehearsed new vocabulary by speaking only" (p. 121). In addition, the study exhibited the ability of music to enhance the learning environment.

Schunk reported that students came to the class with a positive attitude and stayed on-task and engaged throughout singing activities. The music created an overall playful mood and took the drudgery out of text rehearsals.

Later, in 2005, Kennedy conducted a study with middle-school ESL students, again all of Latino background. "Music and movement, active music listening, group chanting and singing, musical games, rhythmic training, music and sign language, and lyric analysis and rewrite activities" were used (Kennedy, 2005, p. 244). The results showed that the experimental group, which was taught with music therapy techniques, did significantly better in story retelling than the control group, which was taught in the traditional way.

In 2008, Kennedy extended his research in a study with the use of music therapy for ESL with Kindergarteners all of Latino background. "Singing, chanting, action songs, music listening and musical games were used to develop pronunciation, vocabulary, story retelling skills, and critical listening skills" (Kennedy, 2008, p. 96) In summary, the study showed that music therapy as the principal teaching strategy for Kindergarten ESL students enhanced their practice and acquisition of English language information (Kennedy, 2008).

Kennedy was particularly inspired by Carolyn Graham, who was not actually a music therapist, but rather a musician and language teacher. She developed jazz chants, which she defined in her book as "the rhythmic expression of Standard American English as it occurs in situational contexts" (Graham, 1978, p. 9). In her teaching, it was found that jazz chants work well with English because it is based on the rhythms of jazz, which has traditionally been composed by native English speakers. In addition, the rhythms that make jazz so characteristic came with slaves from Africa.

Thus, jazz chants could potentially be useful for select African languages and other languages and dialects born out of African languages, such as Jamaican Patois or Haitian Creole. For example, Cassidy (1961) comments on the residual Africanisms in Jamaican Patois, "The slaves ... kept up many African customs ... These activities naturally tended to preserve some African words and patterns of speech, if no actual African language as such" (p.19). Thus, African rhythms have had a widespread effect on languages around the world.

Applications of Music Therapy Techniques for SLA

Since music therapy is the application of music to achieve non-musical goals, experts in the field are well-equipped to find ways to use music therapy techniques to enhance language learning, particularly ESL. In addition, research into using music therapy for language acquisition could be an important stepping stone for medical applications including aphasia, traumatic brain injuries, ESL students with learning disabilities, "foreign accent syndrome," and acquired arrythmia (Patel, 2008, p. 175). The following are some applications of music therapy activities for ESL.

Movement activities can be very diverse. Music can be used simply as background accompaniment to energize and boost students' attitudes and readiness to learn. It may also be used as a catalytic stimulus (Unkefer, 2005, p.194). In this case, students synchronize their movements with the words, ideas, directions, or moods presented in the lyrics. Very direct and specific directions could be "to the left, from "Cupid's Shuffle," or "Put your right hand in," from "The Hokey Pokey." Alternatively, students may listen to the lyrics of a song and improvise their own moves and gestures that express the meaning.

Singing can serve a multitude of purposes. Firstly, students gain experience verbalizing in the target language with the group without the pressure of becoming embarrassed when making a mistake. Group singing can be a great energizer and mood booster (Gaston, 1968). In the case of Schunk, singing was a useful tool for building vocabulary in an authentic, unconscious manner (Schunk, 1999). Melodic Intonation Therapy techniques work in this manner as it can be a useful tool for teaching correct pronunciation and intonation (Thaut, 2005).

As shown by experience, chanting can be a very effective and fun technique when used in the ESL classroom. Carolyn Graham's jazz chants are available for use and can also be used as a model for creating new chants appropriate to the material the class is discussing. Furthermore, Graham's jazz chants seem to be related to the music therapy technique Rhythmic Speech Cuing (RSC) (Davis, Gfeller &Thaut, 2005).

One of the greatest challenges with teaching adults English is finding simple but age-appropriate material. Original songwriting on the therapist's part is a great solution to this problem, since the words can be phrased simply but the content can be made interesting for the group. In lessons, songwriting can used as an application and analysis of grammar, which requires thought from the students (Unkefer, 2005). Having students write their own mnemonic songs might be an effective way of helping students remember more (Moore, 2008). This also targets students' writing and speaking abilities, including conversational skills. (Baker & Wigram, 2005) Topics can range from a narration of everyday activities to stories of fiction and fantasy.

Song lyric analysis has many uses as well. For reading/listening comprehension, a student may bring a song that he or she would like to discuss for meaning. It can be adapted for different levels by the depth of discussion and can vary from basic translation and reading comprehension to deeper discussion of the themes, ideas, and social statements that can be found in the song (Grocke, 2007). For grammar with a focus on forms, the therapist can bring or ask students to bring a song that includes many of the target structures that they would like to discuss.

Even imagery activities can have a place in the ESL classroom. For example, an adaptation of unguided music imaging can be used for providing inspiration for freewriting. The therapist would not provide any suggestion to students, except to write about whatever the music brings to mind. The music may paint a generic picture or bring a specific place or thing to mind, such as the theme song to a TV show. The extra-musical associations could thus become an inspiration to write with very little verbal prompting from the teacher (Gaston, 1968). A similar concept is getting students to create a music collage. This would involve putting together a collage of the items spoken about in the song and then explaining to the class why those items were chosen (Grocke, 2007).

Games are also a great way to lighten the mood of a lesson. By nature, it encourages self-expressive language and social interaction, which requires output (Unkefer, 2005). This is an authentic kind of activity, which makes it more meaningful than a strictly prepared and artificial activity. Recreational performances, such as talent shows, can be also be used to encourage social interaction in a similar way (Unkefer, 2005).

Perhaps the easiest of all, music can be extended into students' leisure time in an enjoyable and fulfilling way. Students may be given a playlist of songs to listen to as their homework to reinforce L2 when outside the classroom (Unkefer, 2005). This is an extension of the "din in the head" concept, or involuntary language rehearsal (Krashen, 1983). Music has a tendency to become an "brainworm" (p. 45), which can be used to the benefit of ESL learners as it provides constant, subconscious repetition (Sacks, 2007).

Method

Subjects

The subjects in this study were 38 English language learners enrolled in an ESL program at a community college in the Southeastern part of the United States. Students' L1s included Spanish, Korean, Chinese, Farsi, and Turkish. Students ranged from beginner to intermediate levels of English. They ranged in age from 20-65 years.

Materials

For this study, researchers used a variety of activities and materials. A smartboard/projector and whiteboard was available for use in the classroom, which allowed for PowerPoint Presentations to be displayed over the course of the session. The PowerPoints supplied pictures and lyrics that were sometimes specially formatted to assist with the assimilation of speaking and tapping for emphasis. Students played sticks or tapped on the table or on laps while the therapists played on djembe and guitar or ukulele.

Procedures

This study was conducted at a community college in the Southeastern part of the U.S. The participants were divided into an experimental group (n = 23) and a control group (n = 15), based on existing class distributions. Two 30-minute sessions of ESL music therapy were held per week with the experimental group over the course of 26 weeks. The experimental group was taught using traditional classroom methods that were supplemented by 30 minutes of the music therapy intervention during the regular class time. The traditional classroom instruction was held three times a week for a total of 3 hours of instruction per session. The control group was taught solely using traditional ESL classroom instruction, held three times a week for 3 hours sessions.

The music-assisted instruction was mainly determined by the content currently being discussed by the ESL teacher. For beginner students, content mainly consisted of vocabulary practical for use in the home, school, work and community; grammar; and social studies in preparation for the citizenship exam. Textbooks used included the following: *English No Problem!* By Trish Kerns and Patty Long, *Workplace Skills Builder OPD* by Jayme Adelson-Goldstein, Norma Shapiro, and Marjorie Fuchs, the series *Focus on Grammar: An Integrated Skills Approach* published by Pearson Longman, and *Preparation for Citizenship* published by Steck-Vaughn. Experimenters collaborated with the ESL teachers conducting the class on a regular basis so that activities were closely aligned with the content being covered at the time. The music activities frequently utilized Carolyn Graham's *Jazz Chants* series designed for ESL education. Chants were taken from the volumes *Jazz Chants* (1978), *Small Talk: More Jazz Chants* (1986), *Grammarchants: More Jazz Chants* (1993), *Jazz Chants: Old and New* (2001), and *Children's Jazz Chants: Old and New* (2002). In addition to these, many popular, educational, and therapist-original songs and activities were presented.

Both groups were required to take the Best Literacy Test as a pre-test and post-test for admission in the ESL program. Also, at the beginning and end of the study, the expressive language skills of 17 of the subjects in the experimental group were assessed by their ESL instructor using the Speaking Checklist and Story Retelling Inventory. Both of these assessment tools are taken from Hurley and Tinajero's textbook, *Literacy Assessment of Second Language Learners* (2001). Although there is currently no reliability or validity data available for these tests, Hurley and Tinajero claim that the English-Speaking Checklist and Story Retelling Inventory are a much more effective means of assessing ESL learners' knowledge than traditional standardized tests.

Both the English-Speaking Checklist and the Story Retelling Inventory are designed to acquire performance data related to a language learner's overall synthesis of spoken language. In contrast, standardized tests are only able to measure the learner's knowledge of specific facts or concepts. In addition, the English-Speaking Checklist and Story Retelling Inventory are both capable of assessing ESL students' knowledge without wasting valuable instructional time. The tests are structured so that an observer (independent or otherwise) may collect data at any point during the learning process, informally and unobtrusively. Standardized tests, however, not only require that instructional time be set aside for assessment, but also demand a more formal testing environment which may result in test anxiety and less than optimal performance.

The English-Speaking Checklist contains 15 items, each of which is satisfied by the observer circling "always," "sometimes," or "never." All of the items on the checklist are behaviors that are easily observed during everyday instruction, related to various aspects such as fluency, diction, coherence, organization, and courtesy. For example, the checklist contains items such as "sticks to the topic," "builds support for the subject," "speaks clearly," "takes turns and waits to talk," and "speaks loudly enough for others to hear," among others.

The Story Retelling Inventory is slightly more focused in scope than the English- Speaking Checklist, and features ten items associated with a Likert scale. The observer is asked to rate the language learner's proficiency for each item on a scale from one to four, with one representing "weak," two representing "average," three representing "very good," and four representing "superior." The ten inventory items not only measure story comprehension, but also specific grammar skills and vocabulary usage.

Examples of inventory items measuring story comprehension include "comprehensively retold the story," "comprehended the story line and plot," "understood role of characters," "understood implied and stated meaning," and "comprehended authors' intentions." Other items assess the learner's usage of verbs, adverbs, adjectives, phrases, compound sentences, vocabulary, slang, and colloquial expressions.

Group 1 – Traditional ESL Classroom Methods

The control group received traditional English-as-a Second Language (ESL) instruction. This instruction was based on repetition of vocabulary words, sentences, paragraphs, and short stories that students read from their textbooks. The regular ESL teacher then engaged the students in discussions about the textbook material. Sometimes pictures related to the ESL material were used to enhance the students' engagement and understanding of the language contents.

Group 2 – Music Therapy Techniques for ESL

The experimental group received ESL instruction in the traditional format that was used for the control group except that they received music therapy assisted ESL instruction in the last 30 minutes of their class time. A typical lesson would take place as follows. The therapists often would setup while the teacher concluded her segment of the lesson. They then opened with a hello song during which they would sing the names of the students and encourage them to sing along. Hello songs used included "Hello, Goodbye" by the Beatles and "La Bamba" with modified lyrics for singing hello. Next in the warm-up, clients often sang a song to practice vocabulary and conversation pertaining to the days of the week, months of the year, or the weather.

Following this, clients were handed sticks or encouraged to clap as the therapists spoke through the jazz chants for the day which was displayed in the PowerPoint presentation. These jazz chants were usually taken from Carolyn Graham's "listen-and-sing" technique and used so that clients could hear the rhythm and prosody of the therapist's speech. Depending on students' speech, the therapist would repeat lines, emphasizes stressed syllables in her speech and by drumming along on a djembe. Activities that followed this were typically dependent on the teacher's lesson plans. Songwriting, lyric analysis, and sing-alongs were often used. Songwriting activities ranged from highly structured, fill-in-the-blank formats to abstract songwriting formats in which a theme was used to build a songwriting web from clients' input. The words and ideas in the songwriting web were then formed into sentences with extensive help from the therapists.

Dalcroze techniques were also heavily incorporated in the activities. For example, as an early icebreaker activity, students were assembled into a circle and asked to introduce themselves by name. They were then asked to count the number of syllables in their name. For every syllable, they were told to come up with a movement. The movements were combined to create a composite sign for their name that visually displayed where the stressed syllables fell. The therapists demonstrated what this would look like by showing their own name signs. Once everyone had found a name sign, the group went around the circle calling out the names of their peers while performing the sign that accompanied their name. This use of movement to show sound was used in later activities, such as a movement activity for learning words pertaining to yardwork where subjects came up with movements that sounded like and looked like the word and its meaning.

At the end of the class, subjects were encouraged to sing-along in saying goodbye, at which point they learned different types of farewell greetings. Songs included an original goodbye song, "Na, Na, Na, Na," and "See You Later Alligator." See Appendix 3 for a full list of activities used over the course of the intervention.

Results

English Speaking Checklist

The first research question posed during this study was as follows: Will there be a statistically significant difference in the experimental groups' pre-test to post-test scores on the English-Speaking Checklist? Using the chi square test, 13 out of the 14 items on the checklist yielded statistically significant differences from pre-test to post-test, where $p \leq 0.05$. In fact, the results remain statistically significant even when $p \leq 0.001$. Item #8, "presents in an organized and interesting way," was the only item that did not have statistical significance.

An analysis of the class frequencies of each score – always, sometimes, never – shows where improvement was made from pre-test to post-test. As shown in Table 1.1, for items #3, #5, #10, #11, and #13, more students received "always" than in their pre-test. For item #14, more students were reported as "never" speaking only to those who share the same native language, which in the context of the ESL classroom is a sign of improvement in expressive skills. Thus, these items exhibited the most improvement. Across all items in the Speaking checklist, fewer students received a score of "never," with the exception of item #14 for which fewer students received "always." Items #2, #4, and #6 decreased in scores for "always," but increased in "sometimes" and decreased in "never." Items #7 and #12 remained entirely unchanged across all score types.

Table 1.1 Speaking Checklist

	Frequency Difference from Pre-Test to Post-Test			Inferences
Categories	***Always***	***Sometimes***	***Never***	
1: sticks to topic	↓ Decreased	↑ Increased	↓ Decreased	Although normally a "decreased" score for "Always" would be negative, the reason why many students did not "always" stick to topic was because as their comfortability with the language increased, they more frequently asked questions that took the class discussion on a tangent. Thus, in this case, these are a positive indication of SLA.
2: builds support for subject	↓ Decreased	↑ Increased	↓ Decreased	Fewer students "never" built support for the subject, although fewer "always" did so.
3: speaks clearly	↑ Increased	↓ Decreased	↓ Decreased	More students spoke clearly "always." Fewer students "never" spoke clearly.
4: takes turns and waits to talk	↓ Decreased	↑ Increased	↓ Decreased	Fewer students "never" built support for the subject, although fewer "always" did so.
5: talks so others in the group can hear	↑ Increased	Same	↓ Decreased	More students spoke clearly "always." Fewer students "never" or "sometimes" spoke clearly.
6: speaks smoothly	↓ Decreased	↑ Increased	↓ Decreased	Fewer students "never" built support for the subject, although fewer "always" did so.
7: uses couteous language	Same	Same	Same	No change occurred
9: supports the topical thesis	Same	↑ Increased	↓ Decreased	More students supported the topical thesis "sometimes" and the same number of
10: answers questions effectively	↑ Increased	↓ Decreased	↓ Decreased	More students answered questions effectively "always," and fewer students did so "sometimes" and "never."
11: is comfortable speaking publicly	↑ Increased	↓ Decreased	↓ Decreased	More students were comfortable speaking publicly "always," and fewer students did so
12: maintains listeners' interest	Same	Same	Same	No change occurred
13: volunteers to answer in class	↑ Increased	Same	↓ Decreased	More students volunteered to answer in-class "always," the same for "sometimes,"
14: speaks only to those who share the same native language	↓ Decreased	↓ Decreased	↑ Increased	In the context of the ESL classroom, this was a negative behavior that needed to be extinguished. Thus, the results support the hypothesis.

Table 1.2 Story Retelling Inventory

	Frequency Difference from Pre-Test to Post-Test				Inferences
Categories	Weak (1)	Average (2)	Very good (3)	Superior (4)	
3: understood the major ideas	↑ Increased	↓ Decreased	Same	↑ Increased	The increase in "superior" and "weak" scores suggests two extremes in understanding among students, likely due to the fact that the class was mixed with beginners and intermediate learners
5: brought a background of information to the selection	↓ Decreased	↑ Increased	↓ Decreased	↑ Increased	An increase in "average" and "superior" scores shows favorable progress. Those who were "weak" improved to "average" and those who were "very good" improved to "superior"
6: analyzed and made judgments based upon facts	↓ Decreased	↑ Increased	↓ Decreased	↑ Increased	An increase in "average" and "superior" scores shows favorable progress. Those who were "weak" improved to "average" and those who were "very good" improved to "superior"

Story Retelling Inventory

The second research question posed during this study was as follows: Will there be a statistically significant difference in the experimental groups' pre-test to post-test scores on the Story Retelling Inventory? Using the chi square test, items #3, #5, and #6 exhibit a statistically significant difference from pre-test to post-test, where $p \leq 0.05$. See Table 1.2.

The results showed that for all three items, more students received a score of "superior" by the end of the ESL instruction with music therapy as a supplementary tool. Similarly, for items #5 and #6, more students received a score of "average" and fewer received a score of "weak."

BEST Literacy Test

The third research question posed was as follows: Will there be a statistically significant difference in the gain scores of the experimental group versus the control group? Using a two-tailed t-test, the results were found not to be significant. This was likely due to the small sample size of both the experimental and control groups. More research should be conducted with a larger sample size to receive statistically significant results.

The fourth research question posed during this study was as follows: Will there be a statistically significant difference in the experimental groups scores from pre-test to post-test? Again, using a two-tailed t-test, the results were statistically significant, where $p \leq 0.05$.

Discussion

Challenges and considerations

A challenge that appeared over the course of the study was that the music therapy intervention was supplemental to the teachers' existing lesson plans. These plans included goals and objectives intended to develop mainly understanding of written language rather than oral and aural language. As a technique that generally encourages "sound before sight," music therapy best lends itself towards goals and objectives for the latter, including intonation, rhythm of speech, fluency, pronunciation, and perhaps even accent. With careful and creative development of activities and techniques, music therapy may also be used to improve the written forms of communication.

The power of music therapy as a tool to teach English was thus not fully realized in this pilot study. More research is needed to explore music therapy's potential effect on language acquisition. This may best be achieved by narrowing the pool of participants to having the same L1 and comparable level of mastery of English at the beginning of the intervention. Also, independent variables may be focused to receptive and expressive verbal communication. With this more specified domain, developing goals and objectives, and in turn a curriculum based on music therapy activities, would be much more attainable.

Truancy was also a problem with this class. Many times, students had to leave the class early for various reasons at the time when the music therapists were about to begin their segment of the lesson. Thus, they did not always get the benefit of the intervention. Attitudes towards the music therapy segment varied as well. Some students were very appreciative of the music therapists coming to teach them, smiling and laughing during these segments. Others did not respond as enthusiastically, especially when music therapy was offered at the end of class.

Future Research

Since research into the use of music therapy for ESL is a burgeoning field, there is much left to be explored. Gfeller suggests further research into the effectiveness of familiar versus unfamiliar tunes for musical rehearsal. Kennedy encourages investigating the effects of the sequence of music therapy activities on the amount of student output, participation, and engagement. In addition to these, further examination could be made into (1) the effectiveness of music therapy as a supplemental versus independent ESL teaching method; (2) the effectiveness of pitched versus non-pitched jazz chants; (3) the efficiency of music therapy ESL following a product-oriented curriculum versus a process-oriented curriculum; (4) the efficacy of music therapy ESL with "focus on form" versus "focus on forms" versus a combination; (5) the efficacy of music therapy ESL with isolated form-focused instruction versus integrated form-focused instruction; (6) the efficacy of music therapy ESL with L1-common population versus L1-different population; and (7) the efficacy of music therapy ESL for beginner versus intermediate versus advanced level learners.

Because of the strong overlap, this may be a field of research for which music therapists and language theorists or language teachers may collaborate. One group has a stronger understanding of second language acquisition along with a large body of research, while the other has the application of effective techniques. Most importantly, both parties may realize the value of music for learning. One ESOL teacher concluded:

> There ought to be a good deal of song in all our classrooms ... If the teacher is so hopeless vocally that he cannot hold a melody line – if he cannot pick out a line of notes on piano, harmonica, or recorder – if he has no record player or tape-recorder – then he must still find some way of getting a song out of his class: perhaps even clapping an acting unpaid stripe on the arm of the most musically gifted of his pupils and getting him or her to act as choir master (Hilton, 1974, p. 89).

References

Alvin, J. (1975). *Music therapy*. New York: Basic Books.

Brown, H. D. (1994). *Teaching by principles: An interactive approach to language pedagogy*. Englewood Cliffs, N.J.: Prentice Hall Regents.

Brown, J. M., & Palmer, A. S. (1987). *The listening approach: Methods and materials for applying Krashen's input hypothesis*. White Plains, N.Y.: Longman.

Cassidy. F. G. (1961). *Jamaica Talk: Three Hundred Years of the English Language in Jamaica*. London, U.K.: Macmillan Education.

Chambers, G. N. (1999). *Motivating language learners*. Clevedon; Philadelphia, PA: Multilingual Matters.

Davis, W. B., Gfeller, K.E., Thaut, M. H. (2008). *An Introduction to Music Therapy Theory and Practice*. (Third edition). Silver Spring, MD: American Music Therapy Association, Inc.

Gfeller, K. E. (1983). Musical mnemonics as an aid to retention with normal and learning disabled students. *Journal of Music Therapy, 20,* 179-189.

Grocke, D. E., & Wigram, T. (2007). *Receptive methods in music therapy: Techniques and clinical applications for music therapy clinicians, educators and students*. London; Philadelphia: Jessica Kingsley Publishers.

Hilton, J. B. (1973). *Language teaching: A systems approach*. London: Methuen Educational.

Hodges, D. A., & Sebald, D. C. (2011). *Music in the human experience: An introduction to music psychology*. New York: Routledge.

Honing, H. (2011). *Musical cognition: A science of listening.* Piscataway, N.J.: Transaction Publishers.

Kennedy, J. R. & Scott, A. (2005). A Pilot Study: The Effects of Music Therapy Interventions on Middle School Students' ESL Skills. *Journal Of Music Therapy,* XLII(4), 244-261.

Kennedy, J. R. (2008). Music Therapy as a Supplemental Teaching Strategy for Kindergarten ESL Students. *Music Therapy Perspectives*, 26, 94-98.

Kennedy, J. R. (2014). *Carolyn Graham's Jazz Chants.*

Krashen, S. D., Scarcella, R. C., & Long, M. H. (1982). *Child-adult differences in second language acquisition*. Rowley, MA: Newbury House Publishers.

Krashen, S. D. (1983). The Din in the Head, input, and the language acquisition device. *Foreign Language Annals 16,* 41-44.

Lozanov, G. (1978). *Suggestology and Outlines of Suggestopedy*, Gordon & Breach: New York.

Moore, K. S., Peterson, D. A., O'Shea, G., McIntosh, G. C., & Thaut, M. H. (2008). The effectiveness of music as a mnemonic device on recognition memory for people with multiple sclerosis. *Journal of Music Therapy, 41, 307-327.*

Mossie, K., Woodruff, B., Hill, J., & Zak, L. (2011, November 14). Gabby Giffords: Finding Words Through Song. *ABC News*. Retrieved from http://abcnews.go.com/Health/w_MindBodyNews/gabby-giffords-finding-voice-music-therapy/story?id=14903987

Murphey, T. (1990). The song stuck in my head phenomenon: A melodic Din in the LAD? *System, 18(1),* 53-64.

Murphey, T. (1992). The discourse of pop songs. *TESOL Quarterly, 26,* 770-774. *Passages Journal.* (1986). Philadelphia: Running Press.

Murray, D., & Christian, M.A. (2011). *What English Language Teachers Need to Know, Volume I*. Chs.11 – 12.

Music of the Brain. Films On Demand. Films Media Group, 2009. Web. 28 Nov. 2015.

Patel, A. D. (2008). *Music, language, and the brain*. Oxford; New York: Oxford University Press.

Sacks, O. (2007). Musicophilia: Tales of music and the brain. New York: Alfred A. Knopf.

Thaut, M. (2005). Rhythm, music, and the brain: Scientific foundations and clinical applications. New York: Routledge.

Effects of Crystal Singing Bowl and Voice Sound Healing Meditation on Mood, Stress, and General Well-being: Reporting from Field Experience

Elise Zotos (Eluv)

Abstract

Crystal Singing Bowl and Voice Sound have shown promise in inducing the relaxation response and helping alleviate anxiety and improve well-being. This research shares the clinical experiences that corroborate the results of several studies, which found sound healing with crystal singing bowls including systems such as mindfulness-based meditations to have a significant impact on health. The positive impacts include inducing a relaxation response, helping alleviate anxiety and improving well-being of diabetic patients. While this was an observational study and not a typical scientifically designed study, I received anecdotal statements from my clients attesting to the overall experience and feelings they encountered during and after the sound healing session. From the statements provided, increased relaxation and feelings of increased well-being, and lowered stress were observed in the areas of mental, physical and spiritual benefits.

Introduction

Crystal Singing Bowl and Voice Sound have shown promise in inducing the relaxation response and helping alleviate anxiety and improve well-being. This research shares the clinical experiences that corroborate the results of several studies, which found sound healing with crystal singing bowls including systems such as mindfulness-based meditations to have a significant impact on health. The positive impacts include inducing the relaxation response, helping alleviate anxiety and improving well-being of diabetic patients. While this was an observational study and not a typical scientifically designed study, I received anecdotal statements from my clients attesting to the overall experience and feelings they encountered during and after the sound healing session. From the statements provided, increased relaxation and feelings of increased well-being, and lowered stress were observed in the areas of mental, physical and spiritual benefits.

Having practiced as a Crystal Bowl sound healer for over 20 years, it has come to my realization that each sound healing performance consistently appears to have a two-way therapeutic effect, thus for the attendees listening and for myself as the facilitator/practitioner. In regular concerts and performances worldwide, many attendees have come to me afterward and expressed their inner emotions, which they observed were derived from experiencing the sound healing performance. For instance, One attendee stated that “This was the most relaxed feeling I have experienced” while the other stated that therapeutic sound healing performance left them feeling “nurtured, calm and extremely relaxed.” In a recent clinical experience, I conducted a presentation using ethereal vocals, Crystal Singing Bowls and Angel Harps to offer a unique mystical serenade of relaxing sounds for the participants. As a result of this observational study, below are some of the statements the participants provided. (Pseudonyms have been used to protect the participants’ identities).

Sound Meditation Protocol

Participants were invited to lay down on yoga mats or blankets (In cases of physical restriction or personal preference participants were seated.) Seated participants faced the musical instruments, while participants laying down had their heads closest to the instruments.

Instruments utilized included 4-8 Crystal Singing Bowls, Tibetan Tingasha an Angel Harp, and my voice.

Crystal singing bowls are made from 99.9% pure quartz crystal. The crystal bowls produce tones of long durations with additional harmonics giving it the sound of several long notes that sing together which attributes to the name "singing" bowls. The bowls can be struck or played around the circumference with a suede mallet. I used both striking and circumference playing.

The Angel harp is also known in India as a "Swarmandal", or Indian Harp which is a 21-36 string instrument originating in India, similar to a Zither.

Tibetan Tingsha are small cymbals typically used in prayer, meditation and rituals by Tibetan Buddhist practitioners. The two cymbals are joined together by a leather cord. The cymbals produce a unique long high pitch ringing tone when struck together.

Procedure

To begin the session a short introduction into sound healing was given, including a brief description of each instrument and details of the session structure. Instructions were given to encourage a perspective of becoming the observer, allowing any thoughts or sensations occurring during the session to simply flow in and out, encouraging the practice of mindfulness. Participants were given in a brief guided meditation and instruction to breathe expansively and include the lower abdomen.

The instruments were then introduced, with the Tingshas used to open and close the session. The Crystal Singing Bowls were at first played alone, then the voice was introduced to accompany the bowls. The angel harp was introduced towards the last part accompanied by voice. A closing guided meditative instruction followed, to assist participants to slowly become cognizant of the room and their surroundings and begin gently moving towards a seated position in preparation for a few closing words to conclude the session.

85% of the duration of the 60 minute session, included voice and singing bowls. 4.5% Introduction preparation and 3.5% closing guidance, 7% angel harp and voice.

Participant Statements

1. Ultimate Relaxation experienced

"I experienced a deepened sensation of relaxation very shortly after commencement of the singing bowls, voice and harp." (Laura H)

2. Overworked and over stressed

"I came in with a lot of stress related anxiety and my breathing tends to be more shallow with stress, I also often find myself actually holding my breath during stressful times. Quite early, about five minutes into the sound healing session, I felt my breathing begin to deepen and slow down. My breathing felt smoother and more even than when I arrived prior to the session. The slower deeper breathing made my whole body feel more relaxed, and I felt much calmer after the session. It felt like my mood was different than when I came in, and I actually felt good, which was quite a contrast to when I arrived for the session." (Susan J). *Tachypnea or shallow breathing occurs when one takes more breaths than normal in a given minute.*

3. New to this

"At first I was not really sure what was going to happen, if anything. I have never experienced this combination of sounds and something like this before, so I was curious. I found the integration of Eluv's voice with the crystal singing bowls and angel harps was very soothing and relaxing. I was pleasantly surprised at the end when I realized I fell asleep at some point during the session. To drift off to sleep so effortlessly is unusual for me... The soothing sounds seemed to disengage my mind and all my circulating thoughts to the point where I was so relaxed, I fell asleep. This would be really helpful at night because my mind is often racing with thoughts which makes it hard to fall asleep. It's been an eye opener to the effectiveness of certain sounds being helpful for relaxation and for me to have this as a way to help me get to sleep is extremely helpful." (Sandra G)

4. Was Recommended

"I have not done any type of sound healing meditation before and was not sure what to expect. A friend of mine encouraged me to go and brought me. Overall I really enjoyed this experience of the angel harps, singing bowls and Eluv's voice. It was enjoyable to the point where towards the end of the session something really interesting happened, I felt the pain that I came in with, in my left knee go away. I wasn't expecting to feel anything in particular, so when I moved my knee and the pain was gone, it was a wow moment for me. The swelling also seemed to be much less than when I walked in. Eluv mentioned at the start of the event, that the sound vibrations enter the body various ways including through the skin and bones, and that sonic vibrations can shift matter and energy in our body. I was surprised at the relief I felt, especially coming in with no expectation and leaving with relief of something that was bothering me before I did the sound session. I really appreciate it." (Casey M)

5. Gratitude

"Firstly I would like to say thank you for a wonderful session. I came in feeling depressed and anxious about some things that are happening in my relationship. I was recommended by a practitioner friend of mine to do this with Eluv. I immediately felt comfortable with the sounds floating around me, they actually felt like they were moving through my body as well. One thing I noticed as my mind started to settle down from all the thoughts was that my breathing changed. The sounds and rhythms created a flow that I realized I was following with my breath. I am not a religious person, but there seemed to be a presence of grace within the sound that was very comforting. Once the session was over I had a feeling of calmness that was very apparent. I would say this was the biggest takeaway from the sound healing session with Eluv, as well as I need to do this more often!" (Tim B)

6. Gratitude

"Dearest Eluv, Oh how I loved listening to the beautiful meditation. It is absolutely divine, the best I have ever heard. I will recommend this to every single person I know. I love your beautiful voice and I love each affirmation, prayer and vision. I feel Divine love and wisdom pouring through you and through this amazing CD. I cannot wait to give this to my surgeon, oncologist and others at the hospital. After listening to your words, I felt a deep sense of well-being and a profound sense of hope. I need to see if you have this listed on Amazon yet as I would love to leave a comment (5 golden stars.) "(Nila W)

7. Soulful and Profound

"Eluv's healing music puts me immediately into a blissful state. I feel a release of what I don't need to hold on to anymore and an immediate recharge of what I need. Her beautiful energy and sounds have a healing effect that I resonate so strongly with. She is very gifted and talented. I highly recommend going to one of her healing music concerts. You deserve to treat yourself!" (Sue P.)

8. Sound Healing

"I have experienced Eluv's sound healing which was a delightfully restful and grounding two hours. Eluv's extraordinary vocal, instrumental and channeling skills are a rare gift which I would recommend to anyone seeking direction or simply a gentle release from the chaos of the everyday world." (Diane M.)

9. Angels

"Relaxing into Eluv's music as it cleared my mind, I could hear angels singing. Thank you for your many gifts and talents!" (Mary L.)

10. A Graceful Transport into My Inner Self

"Eluv crafts an extraordinary environment for the inner journey. The music and the vibrations of the bowls invite you into a deeply satisfying and welcoming experience. There is skill at work here! I was gently led into a state of relaxation and calmness, a place where I was aware and yet gone! This is a remarkable gift to yourself, and a wonderful spiritual treat to give to someone else. I give Eluv my highest regard." (Steve R.)

11. Gratitude

"Thank you for all the wonderful thoughts and prayers listen to your cd many times, my doctor was so surprised to see me after surgery at how well I had recovered. I feel your CD helped so much I am listening to it again tonight thank you." (Dave H.)

12. Healing Journey of Sound is Transformative

"I have experienced many different healing modalities over the years and I have to say that Eluv's Healing Journey of Sound is transformative. Her crystal bowls and dulcet voice help you to relax and surrender to the experience...letting go of all the mind chatter and being in the present moment with the music gently pulsating through your cells and ridding the body of negative, old energy. It's a truly healing experience that I look forward to participating again and again." (Jacy S.)

13. Great way to end a stressful week!

"I've been to 5 of Eluv's singing bowl events and I've had a different experience each time, all of them as lovely as she is. This is great for clearing my often cluttered mind. I have met several really cool people at these events. It's easy, relaxing and medicinal. I recommend it for anyone!" (Heidi H.)

Conclusion

This Clinical experience corroborates the study by Goldsby, et al. (2017) which found crystal bow singing including systems such as mindfulness-based meditations, had an impact in inducing the relaxation response and helping alleviate anxiety and improve well-being of diabetic patients. While this was an observational study and not a typical scientifically designed study, I received anecdotal statements from my clients attesting to the overall experience and feelings they encountered during and after the session. From the statements provided, increased relaxation and feelings of increased well-being, and lowered stress were observed in the areas of mental, physical and spiritual benefits.

Reference

Allen L, Shealy, N. (200). An exploration of the effects of toning and quartz crystal bowls on the energetic balance in the body as measured electrically through the acupuncture meridians. *Subtle Energies Energy Med. 16*(2):5–8.

Backe, EM, Seidler A, Latza, U, Rossnagel K, Schumann, B. (2011). The role of psychosocial stress at work for the development of cardiovascular diseases: a systematic review. Int Arch Occup Environ Health, 2011; 85 (1), 67–79.

Barrass, S. (2014). Acoustic sonification of blood pressure in the form of a singing bowl. Paper presented at: Conference on Sonification of Health and Environmental Data (SoniHED), September 2014; York, UK.

Belujon, P, & Grace, A.A. (2011). Hippocampus, amygdala, and stress: interacting systems that affect susceptibility to addiction. *Ann N Y Acad Sci.*, 2011; 1216: 114–121. doi: 10.1111/j.1749-6632.2010.05896.x

Benson, H. (1975). *The Relaxation Response*. New York: Morrow Press.

Burtner, M., Serafin, S. & Topper, D. (2002). Real-time spatial processing and transformations of a singing bowl In Proceedings of the 5th International Conference on Digital Audio Effects (DAFx-02) 2002 Sep 26 (pp. 123–126).

Cahn, B.R., Polich, J. (2006). Meditation states and traits: EEG, ERP, and neuroimaging studies. *Psychol Bull. 132*(2): 180–211.

Cella, D. (1997). *Manual of the Functional Assessment of Chronic Illness Therapy* (FACIT) Scales.

Creath, K. & Schwartz, G. (2004). Measuring effects of music, noise, and healing energy using a seed germination bioassay. *Journal of Alternative Complementary Medicine*. 2004; 10(1), 113–122.

Dusek, J.A., Out, H.H., Wohlhueter, A.L., et al. (2008). Genomic counter-stress changes induced by the relaxation response. *PloS one.3*(7): e2576.

Gaynor, M,L. (1999). *Sounds of healing: A physician reveals the therapeutic power of sound, voice, and music*. New York: Broadway Books.

Goldsby, T.L., Goldsby, M.E., McWalters, M., & Mills, P.J. (2017). *Journal of Evidence Based Complementary Alternative Medicine. 22*(3): 401–406.

Inácio, O., Henrique, L.L., & Antunes, J. (2006). The dynamics of Tibetan singing bowls. *Acta Acustica United with Acustica. 92*(4):637–53.

Le Scouarnec, R.P., Poirier, R,M., Owens, J.E., Gauthier, J., Taylor, A.G., & Foresman, .P.A. (2001). Use of binaural beat tapes for treatment of anxiety: A pilot study of tape preference and outcomes. *Alternative Therapy Health Medicine. 7*(1), 58–63.

Melville, G.W., Chang, D., Colagiuri, B., Marshall, P.W., & Cheema, B.S. (2012). Fifteen minutes of chair-based yoga postures or guided meditation performed in the office can elicit a relaxation response. *Evidence Based Complementary Alternaivet Medicine*. Article ID 501986.

Padmanabhan, R., Hildreth, A.J., & Laws, D. (2005). A prospective, randomized, controlled study examining binaural beat audio and pre-operative anxiety in patients undergoing general anaesthesia for day case surgery. *Anaesthesia*. 60: 874–877.

Park, E.R., Traeger, L., Vranceanu, A.M., Scult, M., Lerner, J.A., Benson, H., et al. (2013). The development of a patient-centered program based on the relaxation response: The Relaxation Response Resiliency Program (3RP). *Psychosomatics, 54*:165–174.

Perry, F.(2014). *Himalyan sound revelations: The complete singing bowl book*. London: Polair Publishing.

Richardson, S., Shaffer, J.A., Falzon, L., Krupka, D., Davidson, K.W., & Edmondson, D. (2012). Meta-analysis of perceived stress and its association with incident coronary heart disease. *JAMA Cardiol, 110* (12), 1711–1716.

Schussel, L., Miller, L. (2013). Best Self Visualization Method with high risk youth. *Journal of Clinical Psychology 69*(8), 836–845.

Shacham, S. (1983). A shortened version of the Profile of Mood States. *Journal of Personal Assessment. 47*(3):305–6.

Snaith, R.P. (2003). The Hospital Anxiety and Depression Scale. *Health Quality of Life Outcomes*. 1:29.

Staufenbiel, S.M., Penninx, B.W.J.H., Spijker, A.T., Elzinga, B.M., & van Rossum, E.F.C. (2013). Hair cortisol, stress exposure, and mental health in humans: *A systematic review. Psychoneuroendocrinology. 38*(8), 1220–1235.

Terwagne, D., Bush, J.W.M. (2011). Tibetan Singing Bowls. *Nonlinearity 24* R51– R66.

Vitaliano, P.P., Scanlan, J.M., Zhang, J., Savage, M.V., Hirsch, I.B., & Siegler, I.C. (2002). A Path Model of Chronic Stress, the Metabolic Syndrome, and Coronary Heart Disease. *Psychosomatic Medicine. 64*(3). 418–435.

Wahbeh, H., Calbrese, C., & Zwickey, H. (2007). Binaural beat technology in humans: A pilot study to assess psychologic and physiologic effects. *Journal of Alternative and Complementary Medicine. 13* (1), 25–32.

Watson, D., Clark, L.A., & Tellegen, A. (1988). Development and validation of brief measures of positive and negative affect: The PANAS scales. *Journal of Personal Social Psychology. 54*:1063–1070.

Wepner, F., Hahne, J., Teichmann, A., Bera-Schmid, G. Hördinger, A. Friedrich, M. (2008). Treatment with crystal singing bowls for chronic spinal pain and chronobiologic activities—a randomized controlled trial [in German]. *Forsch Komplementmed. 5*:130–137.

Plato's Harmonious Man

A Pilot Study: Effects of a Plato Inspired Mindfulness Method versus Auricular Acupuncture for Anxiety

Adina Dabija, L.Ac., Dipl. O.M.

Abstract

The purpose of this study is to investigate the effects of "charms" or spirited words on patients suffering with self-reported anxiety and compare the results with the effects of auricular acupuncture (NADA protocol). According to Plato's dialogue *Charmides*, in order to restore the health of an aspect of the body, a true physician should first attempt to cure the novice's ψυχή (psyche, soul) - the incorporeal, immutable and indestructible aspect of one's consciousness. This can be done, according to Socrates, by the use of "charms" - words of wisdom that create a certain vibration and harmonize conflicting aspects of the soul: the rational, the appetitive and the spirited. Resulting in persuasion, the "charm" will induce self healing of the novice's soul by lessening his subjective suffering and favoring the re-organization of the soul in a harmonious manner. As a result, the "charms" will create the conditions for a silent interior dialogue within the soul which in turn will lead to the re-tuning of the individual to the cosmic harmonious order. Ten adult subjects participated in the study. The subjects were randomly assigned to either the experimental group (n = 5) or the auricular acupuncture group (n=5). A Generalized Anxiety Disorder-7 item (GAD-7) questionnaire was filled by each patient before the session, a week after the session and two weeks after the session. While there were not enough subjects to yield statistically significant results, the data does indicate that the Platonic Inspired Mindfulness Method group made a 42.4% mean decrease in anxiety GAD-7 score from pre-test to post-test, whereas the acupuncture group exhibited only a 25.5% mean decrease of anxiety score.

Introduction

According to American Psychiatric Association, anxiety disorder is a normal reaction to stress (American Psychiatric Association, 2019). Symptoms of anxiety are excessive nervousness, fear, apprehension, and worry. Anxiety disorders include generalized anxiety disorder, panic disorder, obsessive-compulsive disorders, phobias, social anxiety disorder, agoraphobia, post traumatic stress disorders and more. APA estimates that anxiety disorders are the most common of mental disorders and affect nearly 30 percent of adults at some point in their lives.

Anti-anxiety medications can be habit forming and cause unwanted or even dangerous side effects. Many patients choose therapy, exercise, or self-help strategies as their treatment of choice or in conjunction with anti-anxiety medication.

The American Psychological Association's survey "Stress in America" recently reported that 65% of Americans rated work as their top source of stress (American Psychological Association, 2016). Epidemiologically, this is relevant because chronic stress leads to chronically elevated glucocorticoid levels and contributes to the development of conditions such as obesity, cognitive impairment, hypertension, diabetes, lipid imbalance, brain aging, loss of bone minerals, and immune system dysfunction (Karlamangla, 2002).

Platonic view of the soul

The word ψυχή (psyche), appears for the first time in Homer's Iliad (12th-8th century BC), with the meaning "life-substance", or "breath": a dying warrior bleeds out his psyche onto the ground or breathes it out in his last gasp. Later on in Lucius Apuleius Madaurensis's Metamorphoses (2nd century AD) Psyche is a beautiful mortal woman whose love and sacrifice for her immortal lover, God Eros, earned her immortality, granted by Zeus. One of her trials includes descending in the Underworld - a recurrent theme in Greek mythology - symbolizing victory upon the unconscious and irrational parts of the mind. Her acquired immortality is symbolically alluding to the concept of the immortality of the soul in Hellenism and of the soul's deification through the practice of virtues. Psyche becomes the deity of the soul, symbolizing self-search and personal growth through learning, losing, and saving the real love. In the Platonic tradition an intrinsic structure with specific functions is assigned to the psyche - soul. Its activities include governing the body and practicing virtues such as wisdom, courage, temperance and justice to beautify the soul so it can break free from the cycle of rebirths. In his dialogue Phaedrus, Plato describes the soul as composed of three parts: the Intellect or Reason (λογιστικόν - logistikon), the rational part who loves and seeks the truth; the Appetite (ἐπιθυμητικόν - epithymetikon), the part who seeks the accomplishment of worldly desires; and the Spirited part (θυμοειδές - thymoeides) - the part who aspires and wishes. To explain the interaction of these three parts of the soul, Plato uses the image is of the charioteer (Reason) who holds the reins of two horses: a winged immortal horse - representing the spirited part of the soul and a regular mortal horse - the Appetitive part.

Method

Subjects

Ten adults aged 28-47, four men and six women volunteered to participate in the study. The subjects were randomly assigned to either the experimental group (n=5) or the acupuncture group (n=5). The patients did not have specific knowledge of the Plato's writings, but most were familiar with mindfulness, meditation and acupuncture.

Location

Sol Center, a small meditation studio in Port Washington NY.

Procedure

The experimental method: the Platonic Inspired Mindfulness Method

1. Background

The use of words in mindfulness practices

Words have the power to change our behavior and our performance. Psychotherapy and hypnotherapy are based on the power of words to help patients reframe a traumatic experience. According to research based on MRI studies, affirmations create new neural pathways in the brain (Cascio and al., 2016). However, it takes repetition to build a new neural pathway, this is why most mindfulness practices require at least a 18 days commitment, while research shows that it can take anytime between 18 to 254 days to build a new neural pathway. (Lally, P. and al., 2010).

In a study (Newberg and Waldman, 2012) on the power of words, MRI scans showed changes in the brain function while the subject was exposed to negative words. The study found that "a single word has the power to influence the expression of genes that regulate physical and emotional stress". Following the display of the word "no" for less than a second, there was a sudden release of stress-producing hormones and neurotransmitters. Repetition of affirmations can improve academic performance. In another study (Creswell JD and al., 2013), eighty undergraduates indicated their perceived chronic stress over the previous month and were randomly assigned to either a self-affirmation or control condition. They then completed thirty difficult remote associate problem-solving items under time pressure in front of an evaluator. Results showed that self-affirmation improved problem-solving performance and boosted academic achievement in underperforming chronically stressed individuals. Mindfulness is defined by the National Institute of Health as "a mind and body practice of becoming mindful of thoughts, feelings, and sensations and observing them in a nonjudgmental way" (Mindfulness, https://nccih.nih.gov/health/meditation).

The relaxation response is a mind–body state that allows the individual to decrease cognitive and somatic arousal, thus modifying the HPA axis and the autonomic nervous system (Smith, S.M. 2006). Regular meditative practice leads to improved autonomic nervous system regulation (Benson, H. 1975), attention and focus (Brefcynski-Lewis, Lutz, Schaefer, Levinson, Davidson, 2007), social connection (Van Leeuwen, Singer, & Melloni, 2012), emotional regulation (Jazaieri et al., 2014), pain and stress reduction (Khusid & Vythilingam, 2016), and immune and inflammatory function associated with changes in genetic expression (Bhasin et al., 2013). A National

Health Interview Survey revealed that about 20 million U.S. adults used meditation and mindfulness for health purposes in 2007 (National Institute of Health, 2010).

2. Method

The purpose of the method was to induce psychic harmony between the three parts of the soul through practicing philosophy and dialectics as "the greatest music" (Plato), using specific words and then meditating on them.

The session consisted in a three steps procedure:

1 - Reading 12 quotes from the Platonic dialogues aloud by the participant

2 - Guiding the participant towards an inner dialectics process based on the 12 key words selected from the quotes. The subjects were asked to reflect in a non-condemnatory way and write down one idea that comes to mind for each of the key words.

3 - Integration of the key words in the subconscious mind through a guided imagery meditation inspired by Plotinus' writing "On Beauty". Participants were then asked to continue reading the quotes daily for the following two weeks, to reflect on the key words journaling the reflections and to meditate daily for two weeks using the same meditation.

Musical operations within the soul: philosophical basis of the choice and order of the quotations

1. Tuning to Virtue (Arete)

In Platonism, the practice of virtue makes one's soul intrinsically musical, activating the spirited and the intellect parts of the soul with ethos (ethics) and logos (logic). In Phaedo, Socrates confesses that throughout his life he was recurrently visited by the same dream uttering an imperative task to "make music" (Phaedo 61a). He complied to the oneiric imperative by practicing the greatest music: philosophy, which creates harmony within the soul. One of the most important

Platonic ideas is that all things that exist are reflections of an archetype or a "form" that exists in the νόος - nous (cosmic mind). The Cosmos is a precise organized system reflecting a timeless and changeless reality, based on eternal mathematical laws. The practice of virtue brings one in concord with these laws of nature.

Also, the practice of music was considered by pythagoreans and platonists a form of education (paideia) for the soul. However, "the correctness of music resides in its capacity to give pleasure to souls" (Plato, Laws), whereas "colorful music", created for entertainment, is considered by Socrates intolerable and impious. Good music activates the spirited and the intellect parts of the soul. In Republic 412a4–7, Socrates asserts that the qualification of being musical and harmonious at the highest level belongs to the individual who applies music and gymnastics to the soul.

Omnipresent in myths and a crucial element in the practice of Hellenism, music was considered sacred, of divine origin. Orpheus, the founder of Hellenism, descended in the Underworld with his lyre and convinced Hades to return his wife Eurydice to the living realm. God Apollo, the god of medicine, divination and arts is also the god of music, depicted with a lyre in his hand, while Pan is depicted with his pan flute in all representations. In the Republic, Socrates mentions that the most ancient form of education consists of mousikē for the soul and gimnastikē for the body. Music and gymnastics were essential and mandatory parts of the education of any Greek philosopher.

My choice of excerpts in the experimental group includes words and sentences that represent, allude or point towards virtues:

I prefer nothing, unless it is true. **- Euthyphro**

The unexamined life is not worth living. **- Apology**

Wonder is the feeling of a philosopher, and philosophy begins in wonder. **- Theaetetus**

The divine is beauty, wisdom, goodness. **- Phaedrus**

Reconciliation of opposites through harmony - **Symposium**

Would you have been willing to inquire into or learn what you thought you knew? - **Alcibiades**

Temperance is implanted in the soul, and where temperance is, there health is speedily imparted, not only to the head, but to the whole body. - **Charmides**

You can recollect the things you knew before. - **Meno**

For not by art does the poet sing, but by power divine. - **Ion**

The virtue of temperance has more of the nature of harmony and symphony. - **Republic**

For wisdom is chief and leader of the divine class of goods, and next follows temperance; and from the union of these two with courage springs justice, and fourth in the scale of virtue is courage. - **Laws**

(Pause)

2. Harmonizing the three parts of the soul through practicing "the greatest music" (philosophy) and dialectics

Harmonizing the conflicted condition of the human being is the main concern of the Platonic dialogues. Creating harmony in the soul is the most profound "musical" act and the philosopher is a true musician.

In Greek mythology, Harmonia (Ἁρμονία) is the immortal goddess of harmony and concord. She is the daughter of Ares, the god of war, and Aphrodite, the goddess of beauty and love. Their union means, symbolically, that it is beauty and love that can pacify any struggles and conflict to achieve concord, or harmony.

In the Republic, Socrates explains that the divine gift of music and gymnastics has the purpose of helping man to harmonize the two different elements of the psyche – the spirited and the rational – stretching and slackening them in the right measure, through the tension and relaxation. In Gorgias, expressing the fear of being in disagreement with himself, Socrates says: "I truly deem that it would be a better thing to play a lyre out of tune and without harmony, to direct a tuneless choir, or to be at variance or of another opinion with respect to a multitude of people, rather than to be dissonant (asymphonon) and of a different opinion with one person alone, if that one person is myself." Thus, the musicality of the soul consists in the serene coherence and intimate agreement within. The ability to harmonize the three parts of the soul makes an individual "just". In the Republic (442c10–d1) Plato makes use of the notion of symphoňia to indicate the relationship that runs between the three parts of the soul of a temperate individual. When this act of balancing happens, each soul aspect carries out its own role and occupies its own place. Socrates enlarges the perspective from the microcosm of the human soul to the macrocosm of the State. The virtue that permeates the entire State is that which best assimilates itself to harmonia and symphoňia. The perfect state/ soul can be compared to a complex musical structure, in which the relationship between the sounds creates an ideal of harmony.

The primary method of reconciliation of the three parts of the soul in Platonism is dialectics, or the socratic method - the fabric of most Platonic dialogues. Dialectics is the pursuit of truth through reasoning. "Dialectic is the art of discourse by which we either refute or establish some proposition by means of question and answer on the part of the interlocutors." (Diogenes Laeutius - Plato). There is a three beat rhythm in dialectics: it starts with two different points of view (two different notes), followed by negotiation and ending in agreement (harmony).

In the experimental method harmony is achieved through guiding the participants towards an inner dialectics process (self inquiry on 12 key words). The subjects were asked to reflect in a non-condemnatory way and write down one idea that comes to mind for each of the below questions and continue the process of self reflection during the following two weeks, journaling the answers. Some examples of key words and suggested associate inner dialectic inquiry are:

"Truth": What is authentic in my life? Do I know myself? What do I need to find out?

"Wonder"*:* What is wondrous now in my life?

"Reconciliation of the opposites through harmony" - What do I need to reconcile or balance?
(Pause): Can I be silent and listen?

3. Rhythm

Plato refers to rhythm as a calming and balancing remedy : "When a child is too excited for his own good, his mother tempers his emotion by slowly rocking and singing to him. The baby picks up the rhythm of this movement and he relaxes. Thus the mother counteracts the action of one emotion by substituting another emotion for it. A lullaby makes a baby feel all right by making him feel the way he should feel, that is, inducing a calm, balanced movement in him. (Laws, II, VII. 790e-79Ib.)

In the experimental study I create rhythm through the order itself of the dialogues according to the Platonic tradition, as given indicated by Thrasyllus, an astrologer who is the source quoted by the historian Diogenes Laertius in his Life of Plato - nine tetralogies, each containing four dialogues: **1st tetralogy:** *Euthyphro, Socrates' Apology, Crito, Phaedo;* **2nd tetralogy:** *Cratylus, Theatetus, The Sophist, The Statesman* **3rd tetralogy:** *Parmenides, Philebus , Symposium, Phaedrus* **4th tetralogy:** *Alcibiades, Alcibiades II, Hipparchus, Rival lovers;* **5th tetralogy:** *Theages, Charmides, Laches, Lysis;* **6th tetralogy:** *Euthydemus, Protagoras, Gorgias, Meno;* **7th tetralogy:** *Hippias a, Hippias b, Ion, Menexenus;* **8th tetralogy:** *Clitophon, Republic, Timaeus, Critias;* **9th tetralogy:** *Minos, Laws, Epinomis, Letters.* Some commenters (Black, 2013) suggested that the organizing principle of the canon is the yearly order of the Athenian religious festivals centered around different deities.

4. Tuning with the Cosmos and participation in the world's symphony

Ancient Greeks developed the concept of an orderly and harmonious Universe. Pythagoras (5th century BC) was the first to use the word κόσμος (cosmos) assigning it the meaning of "the order of the universe". He saw the physical world as a reflection of eternal numerical and mathematical laws (Iamblichus, 1926). By participating in heavenly harmony, one could induce spiritual harmony in the soul. In Pythagorean thought, number gives rise to proportion, and proportion gives rise to harmony. Pythagoras laid out a system of correspondences between musical notes, spheres and states of the soul. According to this system, the earth is a globe revolving with the other planets around a central fire and the heavenly bodies are separated from one another by intervals corresponding to the harmonic lengths of strings, and the movement of the spheres gives rise to a musical sound - the "harmony of the spheres”. He explained the harmonious arrangement of things as that of bodies in a single, all-inclusive sphere of reality, moving according to a numerical scheme. The fact that music was denominated in exact numerical ratios demonstrated the intelligibility of reality and the existence of a reasoning intelligence behind it. The sacred decade in particular has a cosmic significance in Pythagoreanism: its mystical name, tetraktys ("fourness"), implies 1 + 2 + 3 + 4 = 10; but it can also be thought of as a "perfect triangle”. To Pythagoras, number was a kind of cosmic matter, like the water or air proposed by other philosophers. The emphasis upon numerical proportions, harmony, and order was an important step toward a metaphysics in which the concept of “form” or archetype is the basic reality, developed later by Socrates and Plato. Tapping into the unchanging aspect of reality or “form” can help one to re-tune themselves to the cosmic order. In Timaeus, Plato comes up with a system of harmonic astronomical–musical ratios on which the fabric of the entire Cosmos is built upon. Through harmony and rhythm, the soul endeavors to tune itself, searching for the ideal musicality of its origins and tuning to the symphonic order of existence. Holographically, the harmonia ("fitting together") of the cosmos reflects the intricate order of the fabric of life. For instance the application of the tetraktys to the theory of music revealed a hidden order in the range of sound. In tuning theory, the Pythagoreans identified the arithmetic, geometric, and harmonic means which underlie the musical scale, as well as the perfect consonances of music, which are mathematical ratios: the octave (1:2), the perfect fifth (2:3), and the perfect fourth (3:4). Continuing the Pythagorean tradition, Plato compared a just and well-ordered society to a well-tuned lyre. While each note retains its individuality, all are proportionally linked together in a larger whole to form a musical scale, and all are interdependent in terms of their reliance on one another. (Plato, Republic 443 D–444). Justice is present in any well-functioning organism including society and the soul.

All things are related through whole–part and proportional relationships (analogia), as in an ecosystem. Tapping into the unchanging aspect of reality or "form" can help one to re-tune itself to the cosmic order. Because of this, the classical Pythagorean metaphor likens the universe to a living organism. Plato describes the Pythagorean view well when he describes the cosmos as "one Whole of wholes" and as "a single Living Creature which encompasses all of the living creatures that are within it" (Timaeus 33 A and 30 D).

In the experimental group, participation in this operation and the integration of all the above operations is done through a guided imagery meditation inspired by Plotinus' writing On Beauty: " Let us then receive by cogitation this our sensible world so disposed that every part may remain indeed what it is, but that one thing may mutually reside in another. Let us suppose that all things are collected as much as possible into one, so that each particular object may first present itself to the eyes; as if a sphere should be the exterior boundary, the spectacle of the sun immediately succeeding, and an image of the other stars and the earth, the sea and all animals should appear within, as in a diaphanous globe: and lastly let us conceive that it is possible to behold all things in each. Let there be then in the soul a lucid imagination of a sphere, containing all things in its transparent receptacle; whether they are agitated, or at rest; or partly mutable, and partly stable. And in this soul sphere we will be repeating each word that was previously said and let it become part of the soul's world:

Let us invoke *the Truth* and let it become the Sun of this sphere...

Let us invoke *a life worth living* and let it become the air, breath it in...

Let us invoke the feeling of *wonder*, and let it become a gentle breeze...

Let us invoke *the beauty, the wisdom and the goodness* and let them become stars...

Let us invoke a sense of *harmony*, and let it become rain...

Let us invoke the ability to *learn*, let it become a fertile soil...

Let us invoke the virtue of t*emperance*, let it become rivers changing into clouds...

Let us invoke the ability to *recollect* all you need - let it become the ocean...

Let us invoke *the power divine*: let it become the North Star

Let us invoke the sense of *a harmonious symphony*: let it become a garden's fragrance...

Let us invoke the *wisdom*, let it become a cave in a mountain...

Let us invoke *silence*, let it become the ineffable night...

(Pause) Let us now contemplate the image of each thing holding the other... the sun, the air, the gentle breeze, the rain, the fertile soil, rivers changing into clouds, the ocean, the north star, the flowers fragrance, the cave and the ineffable night. It is indeed a universe, a whole. But this Divine World is so truly great that its parts become infinite. For where can any thing be said to exist, with which it is not extended? This sensible world, too, is great, and all powers are contained in its ample bosom: but it would be much greater, and that in a manner perfectly ineffable (Pause).

5. Silence

In the experimental study, the Pause is an invitation to a minute of listening to oneself in silence. Silence is an invitation to a journey within. It points to things unspeakable and to a deep listening. "Socrates, are you listening to Socrates?" (Statesman).

Notorious for recognizing the superiority of oral to written teaching - as Socrates proclaims in Phaedrus - the Hellenic tradition has been transmitted from the times of Pythagoras to those of Socrates and Plato in the appreciation of a certain vibratory resonance carried by the voice and the presence of the teacher. The knowledge was not imparted to anyone and even the initiate would have to undergo periods of cathartic silence. In Symposium, Alcibiades declares that Socrates is a Silenus figure, with an outer casing that conceals divine images inside (Symposium 215la4-c3, 216c7-217a1). In his Seventh Letter, Plato denies that Dionysius II and other dubious individuals could have access to that about which he is serious,"for there is no writing of mine about these things, nor will there ever be. For it is in no way a spoken thing like other lessons" (Letters, 341 c1-6).

II. Acupuncture

Acupuncture is a system of medical treatment consisting in applying small needles or pressure to specific points in the body. In this system, health stems from the alignment of qi which means "vital energy" and encompasses the balancing of the yin/yang principles that flow through all of the physical worlds. Studies show that acupuncture can be an effective treatment for chronic anxiety disorders (Wang, 2001). A recent evidence-based synthesis review reported that acupuncture therapy positively affected overall wellness, and mental health (Hempel, 2014).

Since 1972, when it was created, the NADA-standardized 3- to 5-point ear acupuncture protocol treatment was helpful as a stress reduction technique. It has been used for substance abuse problems and coping with trauma. It is successfully used in detox clinics, trauma and disaster relief settings around the world and US army. Points include: "Shen Men," "Sympathetic" (or "Autonomic"), "Kidney," "Liver," and "Lung".

Assessment

Data was collected prior to and following treatment using Generalized Anxiety Disorder-7 item (GAD-7) questionnaire. The patients filled in the GAD-7 following a week and two weeks after the procedure.

GAD-7

Over the last 2 weeks, how often have you been bothered by the following problems?	Not at all	Several days	More than half the days	Nearly every day
1. Feeling nervous, anxious or on edge	0	1	2	3
2. Not being able to stop or control worrying	0	1	2	3
3. Worrying too much about different things	0	1	2	3
4. Trouble relaxing	0	1	2	3
5. Being so restless that it is hard to sit still	0	1	2	3
6. Becoming easily annoyed or irritable	0	1	2	3
7. Feeling afraid as if something awful might happen	0	1	2	3

Total Score ___ = Add Columns ___ + ___ + ___

If you checked off any problems, how difficult have these problems made it for you to do your work, take care of things at home, or get along with other people?

Not difficult at all	Somewhat difficult	Very difficult	Extremely difficult
☐	☐	☐	☐

(GAD-7) questionnaire. The patients filled in the GAD-7 following a week and two weeks after the procedure.

Results

Acupuncture group

	GAD-7 prior to tx	1 week after tx	2 weeks after tx
A.F.	5	3	5
A.K.	10	6	8
G.G.	12	7	10
D.J.	11	10	10
E.G.	9	5	6
Total	47	31	39
		16 34% decrease	8 17% decrease

Experimental group

	GAD-7 prior to tx	1 week after tx	2 weeks after tx
S.I.	11	11	9
C.V.	6	5	3
E.R.	10	8	2
H.G.	9	8	6
C.C.	7	5	4
Total	53	37	24
		16 30% decrease	29 54.8% decrease

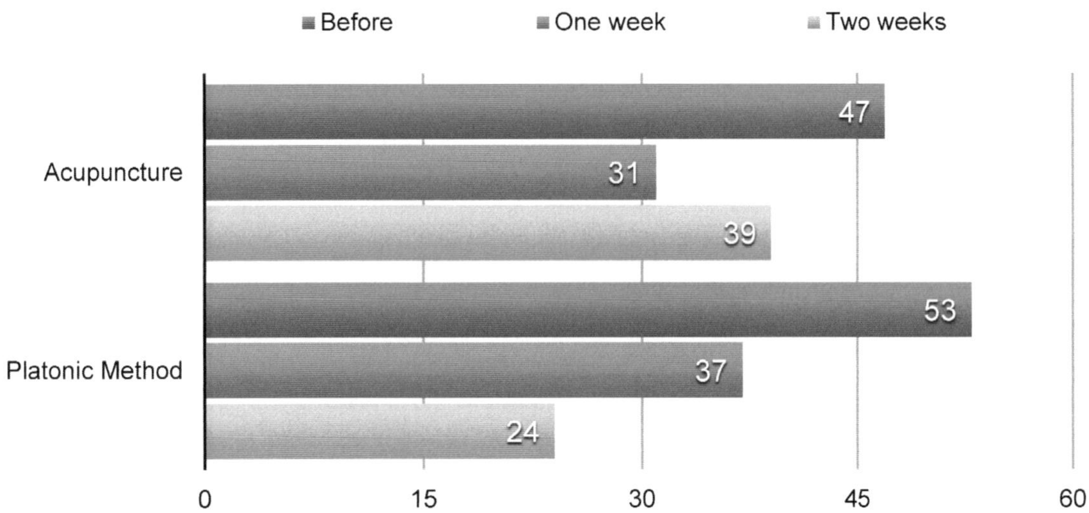

Discussion

Although acupuncture resulted in a bigger (34%) decrease of the anxiety scores than the mindfulness group (30% decrease) a week after the treatment, the rapport reversed two weeks afterwards: the anxiety scores continued to decrease in the mindfulness group, showing 54.8% decrease, while in the acupuncture group they showed a clear increase tendency towards the initial values (17%), even that most participants were sent home with ear seeds for continuous stimulation. The participatory experience offered within the Platonic Inspired Mindfulness Method, emphasized by the loud reading repetition of the quotes on a daily basis for two weeks overweights the effects of auricular acupuncture.

The limitations of the study include the small size of the groups, no double blinding, previous exposure to meditation and lack of control of "homework" (journaling, ear seeds maintenance). A pre-existing commitment to a religious path may be a limiting factor for future participants, as well a pre-existent belief in bottom-up approach ("I need medication").

Health restored due self-reflection and cultivation of virtue is different health that has been restored by passive participation in a medical procedure, even a holistic one. In Platonic view, virtue restores the health of the soul, which in turn restores the health of the body. Cultivation of virtue in the experimental study is done by connecting with "the world of forms", or archetypes, represented by the key words (truth, wonder, beauty, goodness, harmony, wisdom etc), which act as correctors of the anxiety generating patterns of thought. How one relates himself in illness to a central model of perfection offers a priceless opportunity to tap into one's authentic nature and ultimately improve one's spiritual well-being.

References

American Psychiatric Association, (2019). https://www.psychiatry.org/patient s-families/anxiety-disorders/what-are-anxiety-disorders.

American Psychological Association. (2016) *Stress in America: The impact of discrimination*. Stress in America Survey http://www.apa.org/.

Benson H., Greenwood M.M., & Klemchuk, H. (1975) *The relaxation response. Psychophysiologic aspects and clinical applications*. International Journal of Psychiatry in Medicine, 6, 87-98.

Bhasin MK, Dusek JA, Chang BH, Joseph MG, Denninger JW, et al. (2013) *Relaxation Response Induces Temporal Transcriptome Changes in Energy Metabolism, Insulin Secretion and Inflammatory Pathways.* PLOS ONE 8(5): e62817. https://doi.org/10.1371/journal.poe0062817.

Black, Harper McApline, *The Canon of Plato* (Thrasyllus' Attestation), August 17th 2013, http://harpermcalpineblack.blogspot.com/2013/08/the-canon-of-plato-thrasyllus.html

Brefcynski-Lewis, J.A., Lutz, A., Schaefer, H.S. Levinson D.B. & Davidson, R.J. (2007). *Neural correlates of attentional expertise in long-termmeditation practitioners*. Proceedings of the National Academy of Sciences, 104, 11483-11488.

Cascio, C. N., O'Donnell, M. B., Tinney, F. J., Lieberman, M. D., Taylor, S. D., Strecher, V. J., & Falk, E. B. (2016). *Self-Affirmation Activates Brain Systems Associated with Self-Related Processing and Reward and is Reinforced by Future Orientation*. Social Cognitive and Affective Neuroscience, 11 (4), 621-629. https://doi.org/10.1093/scan/nsv13 6

Creswell JD, Dutcher JM, Klein WMP, Harris PR, Levine JM, (2013) *Self-Affirmation Improves Problem-Solving under Stress.* PLoS ONE 8(5): e62593. https://doi.org/10.1371/journal.pone.0062593

Hempel, S, Taylor SL, Solloway MR, Miake-Lye IM, Beroes JM, Shanman R, Booth MJ, Siroka AM, & Shekelle PG (2014a). *Evidence map of acupuncture.* Evidence-Based Synthesis Program Reports. Washington, DC: Department of Veterans Affairs.

Iamblichus, Thomas & Taylor, (1926). *Iamblichus' Life of Pythagoras or, Pythagoric Life.* J.M. Watkins.

Jazaieri, H., McGonigal, K., Jinpa T., Doty, J., Gross J., Goldin, P.. (2013). *A randomized controlled trial of compassion cultivation training: Effects on mindfulness, affect, and emotion regulation*, doi 10.1007/s11031-013-9368-z.

Karlamangla, A. S., Singer, B. H., McEwen, B. S., Rowe, J. W., & Seeman, T. E. *Allostatic load as a predictor of functional decline.* MacArthur studies of successful aging. J Clin Epidemiol. 2002 Jul;55(7):696-710.

Khusid, M. A., & Vythilingam, M. (2016). *The emerging role of mindfulness meditation as effective self-management strategy, Part 2: Clinical implications for chronic pain, substance misuse, and insomnia.* Military Medicine, 181(9), 969-975.

http://dx.doi.org/10.7205/MILMED-D-14-0678

Lally, P. van Jaarsveld, C. H., Potts, H. W. and Wardle, J. (2010), *How are habits formed: Modelling habit formation in the real world*. Eur. J. Soc. Psychol., 40: 998-1009. doi:10.1002/ejsp.674.

Plato, *Charmides,* Jowett 1892: 157.

Wang, Shu-Ming, MD, and; Kain, Zeev N., MD. *Auricular Acupuncture: A Potential Treatment for Anxiety. Anesthesia & Analgesia*: February 2001 - Volume 92 - Issue 2 - p 548– doi: 10.1213/00000539-200102000-00049.

National Institute of Health, *Exploring the Power of Meditation*, June 25th, 2010, https://nccih.nih.gov/about/offices/od/2010-06.htm.

Newberg, A.B. and Waldman, M.R., *Words Can Change Your Brain: 12 Conversation Strategies to Build Trust, Resolve Conflict, and Increase Intimacy* (2012), New York, NY: Hudson Street Press, 288 pp.

Smith, S. M., & Vale, W. W. (2006). *The role of the hypothalamic-pituitary-adrenal axis in neuroendocrine responses to stress.* Dialogues in clinical neuroscience, 8(4), 383–395.

Van Leeuwen, S., Singer W., & Melloni L. (2012). *Meditation increases the depth of information processing and improves the allocation of attention in space*. Frontiers in Human Neuroscience, 6, 1-16.

Wang S M , Kain ZN. (2001) *Auricular acupuncture: a potential treatment for anxiety*. Anesthesia Analgesia ;92(2):548-53.

Sacred Sounds of the Chinese Tradition

Yuanming Zhang

Abstract

This workshop will introduce the traditional Chinese methods of internally resonant sounds known as zhou (咒) and cosmically resonant sounds known as nian (念). The internally resonant sounds are considered to spontaneously emit from the organs, meridians, and dantian (丹田centers of power). Through chanting these sounds, the traditional Chinese believed we can improve the strength and functions of our bodies and promote health in ourselves and others. Further, the use of the cosmically resonant sounds is considered to improve the more rarefied aspects of our minds and spirit and help in the elevation of consciousness toward spiritual development. These techniques are an example of the long tradition of Chinese medical methods that utilized sound as healing. Broadly called zhuyou (祝由) these methods have been documented in texts that date back several thousands of years such as the Yellow Thearch's Inner Canon (Huangdi Neijing 黄帝内经). This workshop will present a method of using these sounds for healing and personal development that can be integrated with other therapeutic modalities. No prior experience with this method is required to learn practical skills for clinical application.

Note: This article is based on the experiential workshop conducted at the 2019 ISQRMM Conference

Introduction

Sound is part of man's innate nature. In the primitive period, humans used their voices for communication but also for song and music to ease their suffering, express their joys, or commemorate their experiences. In the Qingcheng Mountain Tradition of Qigong Self Cultivation, we study how sound is both a natural expression of the human body, as well as a tool that can be refined to promote health of the body and mind either as health maintenance or as medicine.

Resonances from Inside the Body

The traditional Chinese belief is that everything makes sound since everything vibrates and moves. The concept of Qi (气) or the energy of the body implies that the body can warm itself and regenerate itself thereby healing or maintaining health. Warmth is motion; just as you rub your hands together and they become warm. The Chinese have long believed that the breath stokes the movement of the organs through stimulating the Qi energy. We call that resonance from the organs a zhou (咒). The word zhou is made of two boxes representing the human mouth and a wave like line representing a spontaneous emanation of sound. Just like when you pick up something heavy and you grunt from the force, so we believe that the organs put out a spontaneous sound when they are strongly acting in the metabolism. Ancient scholars and doctors from the Qingcheng Mountain studied the body and recorded five sounds that are related to the five major organs of the body. They then refined the practice of these sounds so that a person could mimic the sounds of the health organs and thereby maintain or regain health. Here is a list of the five major organs and the sounds that are considered to make them healthy:

Organ	**Zhou Resonance**
Lung	Ah 啊
Kidney	Hai 咍
Liver	Si 咝
Heart	Ha 哈
Spleen	Oh 哦

So it is possible to make these sounds and improve the health of the organs. But just making the sounds with your mouth is not enough. You need to "play" your body as if it is an instrument and vibrate the sounds inside so that the organs are stimulated by these sounds. This is the secret to vibrant health. You use your body as a great wind instrument and play these fundamental sounds so that they shake your core. By doing so you can strengthen your core organic functions and attain better health. Also this ancient practice forces everyone to breath deep and oxygenate the blood invigorating the circulation to reach every cell. The order of the sounds can follow certain Traditional Chinese Medical Theories, but they can also be used free form to simply play the intrinsic tune that lies deep within the being of every person imbued by the divine. So playing a self generated tune of these organ sounds is the ultimate natural song of the human body. It brings you back to childhood where you simply sang, or hummed a tune of your own devising that fully satisfied you. We consider that to be the Daoist virtue of ziran (自然) which is 'naturalness' or just being who you are. It liberates you from expectation and allows you to simply act without a set mode of what is 'good' or 'right.' We consider that as an expression of another Daoist idea, that of wuwei (无为) or 'non-action' which means not acting as your think you are supposed to, but rather just acting without self criticism or embarrassment. Just play the music of yourself through yourself! So that's the first technique I want to share with you in this workshop.

Self Healing Through the Power of "Ouch"

In addition to the sounds of the organs, in the Qingcheng Mountain Tradition, the ancient scholars observed the way people responded when they were injured or sick. People make a sound when they suffer that sounds like, "a-yo" (啊唷). Well we consider that to be a sound that is coming from the body trying to heal itself, like the sound of an engine trying to turn over in a car. So the sages of this tradition took that sound and made it into a practice. We resonance that sound "a-yo" softly at first and then stronger and stronger until it becomes rousing. We even do it with a marching step to invigorate the body and move the Qi energy throughout the channels that are used for acupuncture. This simple practice is the best way to help with chronic pains and concerns. It uses the breath, movement, increasing resonance that invigorates the circulation, and shifts the focus from the pain condition to a transcendent one which expands the spirit. The other wonderful thing is that this method is free. It requires no medicines that could have side effects, and all the benefits are created by the natural changes in body metabolism. Although these methods have not had many controlled studies done to support them, we have the weight of a tradition that goes back well over a thousand years, as well s the clinical observation of lineage doctors of traditional medicine who both developed these methods and continue to employ them to this day.

Cosmicization of the Self

The last part of today's presentation relates to what we call nian (念) or cosmic resonances. The word nian is also used to mean chanting. But these sounds are considered to relate to larger cosmic forces similar to the Indian idea of a mantra. Although there are many of them which relate to the Chinese traditional view of the cosmology, we will only focus on one called hong (哄) which is the sound that correlates to the integration of the individual to the universe. So this sound is the byproduct of you existing as an integral part of the working of the world. By resonating this sound we can attune ourselves to the cosmos and become more at peace. So we often do this sound after dancing or marching while strengthening our organs as we described above. Then as we build into a crescendo of sound we switch to the cosmic resonance hong. By doing so we dissolve our border between the world and ourselves thereby realizing our interconnection to all things. If we can attain this cosmicized state, why would we be depressed or upset? Realizing our integrated divine nature is the highest of medicines.

Conclusion

I hope that this workshop has been useful as an example of a living tradition of sound and music being used as medicine. The twenty-first century has given us many wonders in technology and medicine. However, traditional methods have the weight of empiricism behind them and allow us to access the intrinsic medicine that lies within ourselves.

Contributors

David O. Akombo, Ph.D. is Associate Professor and Faculty Fellow in the College of liberal Arts at Jackson State University. Before joining Jackson State University, Dr. Akombo was a Diversity Fellow at Weber State University in Utah. Intrigued by the peculiarities of music and its interdisciplinary confluence with medical and psychological practices, Dr. Akombo has studied and researched the effects of music on children and young adults both in schools and in hospitals. As scholar and performer, Dr. Akombo balances his research interests amicably as a music educator, ethnomusicologist, composer and singer/drummer, having worked in Africa, The United States, and Southeast Asia where he studied with Balinese artists. His books, Music and Healing Across Cultures (Ames, Iowa: Culicidae Press, 2006) and Music and Medicine: Connections Found (Seaburn, 2009) unfold the mechanics of the relationships between music, culture, physiological and spiritual states in people from the anthropological and epistemological inquiry regarding music and healing. Dr. Akombo has worked jointly with the University of Florida's Center for Arts in Healthcare Research and Education (CAHRE) with which he collaborated to establish the Arts in Medicine (AIM) program in Nairobi, Kenya. Located at Mater Hospital and funded by the Legislature and State of Florida Center for Cultural Affairs in Tallahassee, this is the first known Arts in Medicine program in Africa. Dr. Akombo continues to be active in music education research, and interdisciplinary scholarship having conducted clinics in several countries. He is a 2017 recipient of the NSF Research Training Scholarship. He is member of the Music Educators National Conference, The national Social Science Association and co-founder and President of ISQRMM. He can be contacted at dakombo@isqrmm.com

Deirdre Cogan, ATR-BC, LPC, ATCS, CCTP, has 30 years of clinical experience in the public and private sectors. She began her tenure with the District of Columbia government creating models to address the complex, varied needs of intercity youth in the custody of the Child Welfare System and Department of Youth Rehabilitation Service. She currently oversees the Creative Arts Therapies clinical and training programs at Saint Elizabeth's Hospital in Washington, D.C. She is a consultant to the Center for Post Traumatic Disorders, an internationally recognized program that offers treatment to adults with dissociative and trauma-based conditions. Deirdre's research interests are in designing programs that address recovery challenges facing forensically-committed women who face barriers due to poverty, gender-inequality and cultural stigma. She has received several meritorious recognitions from the DC Department of Behavioral Health and she is the recipient of the American Art Therapy Association's "2015 Clinician of the Year Award." She can be contacted at deirdre.cogan@dc.gov

Laryssa M. Creswell, EdD, MT-BC, LCPC, earned her Doctor of Education in Counseling Psychology from Argosy University/Washington DC. She is a board-certified music therapist and a licensed clinical professional counselor who has worked in the field of mental health in various settings for the past 17 years. She has worked as a music therapist with children, adolescent, adult and geriatric populations. Dr. Creswell has devoted many years working collaboratively with other creative arts disciplines to enhance treatment for those with severe persistent mental illness. She has a wealth of knowledge in program development and has a long-standing interest in developing and modifying treatment provisions for women with severe and persistent mental illness, trauma, and substance use disorders. She has published research on women's perspectives of treatment in the psychiatric institution and she is invested in creating a robust gender-responsive program for women with co-occurring disorders. She can be contacted at laryssa.creswell@gmail.com

Derrick Crow holds a PhD in Education with a concentration in Arts Integration and Curricular Theory. He also holds a Bachelor's of Music Education from Eastern Illinois University, and a Master's of Music from Southern Illinois University Carbondale. His research interests include music medicine, aesthetic education, and arts integration. Through research and pragmatic experiences, he is striving to create educational growth that encourages creative and intellectual exploration. He has had the privilege to work as a music educator at the primary, secondary and collegiate levels, has worked as a student teacher supervisor for the Master's of Arts in Teaching Program, and as an adjunct lecturer at Southern Illinois University Carbondale. He is currently practicing his skills as an educator and musician by teaching young musicians in Danville School District 118. He can be contacted at crow.da19@gmail.com

Adina Dabija, L.Ac., MSTOM, Dipl. O.M. Adina is the Founder Sol Center www.sol.center Adina Dabija is a writer, philosopher and western esoteric arts practitioner. Adina currently lives in New York, where she practices Chinese medicine. She published several books of poetry and fiction. She is also a priestess of Dionysus in the ancient Orphic mystery tradition, an initiatic lineage of Hellenic priesthood passed down directly from teacher to student. She can be contacted at imilovan@unipu.hr

Dr. Iva Milovan Delic, Ph.D. Assistant professor on History Department of Juraj Dobrila University of Pula. In February 2013 she defended her doctoral dissertation (Spanish Flu 1918/1919 in the City of Pula – Social Aspects of the Spanish-Flu Mortality and Urban Anatomy of Pandemic) at the Faculty of Humanities and Social Sciences, University of Zagreb, thus finishing Doctoral studies of modern and contemporary Croatian history in the European and global context. From June 2017 she holds the position of assistant professor at the History Department and teaches several courses regarding contemporary history, one of which is a History of diseases in 19th and 20th century Europe on graduate studies of history. Her main scientific interest is history of health in the first half of the 20th century. She published several professional and scholarly papers and participated in conferences at home and abroad. She can be contacted at imilovan@unipu.hr

Lora Heller, MS, LCAT, MT-BC is Special Projects Coordinator and adjunct instructor for Music Therapy at Molloy College. Lora has worked as a music therapist in diverse settings including geriatric care and adolescent psychiatry in addition to early childhood music therapy in both medical and educational settings. After completing her Master of Science in Special/Deaf Education, she developed Baby Fingers LLC - a music and sign language program for families with young children. Lora has supervised and trained music therapy students from Molloy College and elsewhere since 1994 and has presented at conferences both regionally and nationally. The author of seven sign language books for children, she also co-authored a chapter in Meeting the Needs of Our Clients Creatively, a college textbook about music therapy with the dying and bereaved. She can be contacted at lora@mybabyfingers.com

Kristopher Hong is a talented classical guitarist that is employed as a newly hired, full time music therapist, in the Atlanta Public Schools. He can be contacted through his co-author Roy Kennedy at rkennedy@uga.edu

Ricardo Hurtado is a board certified and licensed music therapist graduated from the University of Georgia with a Bachelor's Degree in Music Therapy. Ricardo Hurtado has been working with children in music education and performance settings for more than 25 years. Hurtado has served as the school director, music teacher/instructor and children choir director for many organizations including Education Through Music-LA, Los Niño's Primero, Sandy Springs Parks & Recreation, Sandy Springs Mission, Inc., Action Ministries Atlanta, School of Music and Arts Fountain of Love United Methodist Church, and The Salvation Army Doraville. Along with Hurtado's long-term commitment to helping students discover and develop their artistic music skills, Hurtado has been particularly passionate in discovering new and innovated ways to combined his professional music therapy skills and his music education experience to develop low-budget music programs specially design for non-profit organization in the low-income family sectors. In addition, Hurtado has developed the Cyber Orchestra Method which employs a group-based music program design strategically to allow a combination of music therapy approach along with the standard music education pedagogy. The Cyber Orchestra has been fully implemented in different organizations including Los Niño's Primero and Actions Ministries. Ricardo Hurtado continues to develop in his professional carrier and remains serving and working for the public educational system. Furthermore, Hurtado is currently in the process of starting his own music therapy private practice in Los Angeles, California. For more information, please contact Ricardo Hurtado at: hurtado_ridardo@hotmail.com.

Samantha Jones is a talented Music Educator/Researcher that has recently graduated with a Masters in Music Education from the Hugh Hodgson School of Music. She can be contacted through her co-author Roy Kennedy at rkennedy@uga.edu

Carla Rose Kelly is a Global Music Award winner, 2019, for Healing/Meditation Music for her composition PEACE (https://youtu.be/YKpXeHDo1uY). She taught Royal Conservatory piano & theory for 25 years, while becoming involved in Music & Medicine research and developing her own private practice in Sound Therapy. She has had the privilege of speaking internationally about her published research, hosting workshops and classes in Japan, England, Norway, Canada and USA. She is the patented inventor of new sound healing instruments and methodology to heal. Her Certification Program in Sound Therapy includes 3 levels of training and is based on bridging the Science of Sound with Spirituality, the latest research in Music & Medicine, Psychology, Neuro-science, Music & Math. It encompasses a beautiful connection of how profoundly powerful music and sound can be for healing and ascension. The program specializes in using the SACRED SCALE ®, 9 healing frequencies, 174-936hz which were recently decoded from the Bible. She is playwright, co-composer and producer of the 2018 musical, Let the Music Come Alive. It is a new Canadian musical telling the tale of how a young girl uses music to help her father's heart heal. This musical hosted two Global Music Award winning songs: I'm Alone, 2018 for new Musical Theater Music, and her composition, Peace. info@carlarosekelly.com

Roy Kennedy, Ph.D., MT-BC, LPMT is an associate professor of music therapy at the University of Georgia, USA. While previously providing music therapy services to various populations, his primary research interests include guitar pedagogy for music therapists, music therapy for English-as-Second Language learners, adults with substance abuse problems, and skill generalization of students with developmental disabilities. Dr. Kennedy has presented his work in the USA, Brazil, South Korea, and Costa Rica. He continues having an interest in the international development of the profession. He can be contacted at rkennedy@uga.edu

Lillie Klein, MS, MT-BC recently received her Masters in Music Therapy from Molloy College. She completed her internship at The Louis Armstrong Center for Music and Medicine at Mount Sinai Hospital, where her focus was in adult inpatient care and oncology. She currently works as a music therapist for Hospice of New York, providing end-of-life care for patients. Lillie is also in affiliation with Long Island Music Therapy Services, where she works with a range of ages and populations. In addition, Lillie proudly works with Lincoln Center Moments, a performance-based program for individuals with dementia and their caregivers. She has participated in several mission trips to Jamaica and during the summer, Lillie volunteers at a bereavement camp for children and adolescents. Prior to her work in music therapy, Lillie earned her Bachelors degree in Musical Theatre and studied with the British Academy of Dramatic Arts at Oxford University. She has presented her work for a variety of interdisciplinary forums, including Mount Sinai Hospital and regional palliative care conferences. She can be contacted at LKlein@lions.molloy.edu

Julia MacKinley is a graduate of the Psychology Program at the University of Western Ontario and pursued a graduate degree at the University of Edinburgh in Human Cognitive Neuropsychology in Canada. In addition, she is a member of the Canadian Psychological Association. Her research started with an interest in posttraumatic stress disorder and how individuals process information about themselves. This research later developed into an interest in the uses of yoga, meditation and music as forms of healing, particularly in the context of psychiatric illness. Currently, Julia works as a psychometrist in adolescent and adult mental health with a focus on rehabilitation and recovery. She can be contacted at julia.mackinley@gmail.com

Aisling Mansion is a talented vocalist who just completed her music therapy internship and she also minored in Teaching English to Speakers of Other Languages (TESOL). She can be contacted through her co-author Roy Kennedy at rkennedy@uga.edu

Jon Reichert currently works as a music therapist at Brunswick Hospital in Amityville, NY, a six-unit acute care psychiatric facility that provides care to adults, geriatrics, and adolescents. He has worked at Brunswick for two years, primarily conducting groups in each of the facility's units. Jon bases his work within a humanistic theoretical orientation, and firmly believes in and practices music as therapy. Prior to Brunswick, Jon did his internship in one of the adult inpatient psychiatric units at Woodhull Medical Center in Bedford-Stuyvesant, Brooklyn. He is an MT-BC and has a limited permit to conduct creative arts therapy in the state of New York. Jon received his MS in Music Therapy from Molloy College in May 2018. Prior to becoming a music therapist, Jon worked for 28 years as a financial analyst in New York City's financial district. He received his MBA from Indiana University in 1986. He can be contacted at jon.d.reichert@gmail.com

Kate Stanley is a board-certified music therapist in Washington, D.C. She received her masters in music therapy from Temple University and has been practicing for eight years. Her other research interests include music and storytelling, insight and schizophrenia, and the use of music and the creative arts in adjudicative competency restoration. She can be contacted at kate.stanley@dc.gov

Dr. Jasna Valic, MD is a specialist Epidemiologist, and Head of the Epidemiology Department Service on the Public Health Institute of Istrian County in Pula, CROATIA. School year 1977/78 attended in Los Altos High School, Hacienda Hights, LA, California, USA. Graduated on University of Zagreb, School of medicine, as Medical Doctor in 1987. Year 1993/94 worked in Tumor Registry of "Lega Italiana per la lotta contro i tumori", Sezione Bologna, Italy. In year 1998 she specialised in epidemiology at University of Zagreb, School of medicine. She has been working at Public Health Institute of Istrian County in Pula since 1994. She has presented scientific and professional papers on several scientific and professional conferences and public lectures regarding not only medicine, but also history of medicine. She can be contacted at epidemiologija@zzjziz.hr

Dr. Sabina Vidulin, Ph.D. is the Head of the Department of Music Pedagogy and leader of musical-pedagogical courses at the Academy of Music in Pula. She is the founder of the International Symposium of Music Pedagogues and the International Forum of Music Pedagogy Students. She has held numerous lectures throughout Croatia and abroad at recognized music academies: from the Salzburg's Mozarteum, Vienna's MDW, the Royal Conservatory in The Hague, to music academies in Germany, Belgium, Poland, Italy, Lithuania, Latvia, Czech Republic, Hungary, Cyprus, Slovenia, Bosnia and Herzegovina and Serbia. She wrote four books, twelve chapters in books, about fifty scientific papers published in Croatia, Bosnia and Herzegovina, Serbia, Slovenia, Italy, Czech Republic, Slovakia, Hungary, Poland and Lithuania. She is the editor of a songbook, five symposia proceedings and one monograph. Vidulin is a member of editorial boards in six international scientific journals, member of the publishing committee of the publisher Helbling (Innsbruck-Esslingen-Bern-Belp) and the national coordinator for the European Association for Music in Schools. She is inovolved in the organization of international symposia in Croatia and abroad as a member of scientific committees. She reviewed books, curricula and scientific papers. She received the State Award Ivan Filipović and the award from the Croatian Society of Music and Dance Pedagogues. She can be contacted at svidulin@unipu.hr

Lauren Viljama, is a University of Georgia alumnus. She completed her music therapy internship and is employed as a full time music therapist at The George Center, where she works with children that have autism. She can be contacted through her co-author Roy Kennedy at rkennedy@uga.edu

Yuanming Zhang comes from a long line of practitioners of the Qingcheng Mountain Tradition of Self Cultivation and Qigong. He has had extensive training in traditional Chinese cultural arts, meditative and breathing practices, martial arts, and Traditional Chinese Medicine. He has taught internationally since the 1990's spreading the traditional wisdom of his native China to the world. He works extensively with the United Nations, and Chengdu University of Traditional Chinese Medicine. He can be reached at masterzhang@yahoo.com

Eluv (Elise Zotos) Specializes in Healing Music and is also Artist, Channel/Medium & Radio Personality. Born and educated in Melbourne, Australia, Eluv is the host of "Ultrasounds" radio show on WMNF, 88.5 FM Tampa. Thursday's 10pm-Midnight. She creates beautiful inspirational music, guided meditations, and visualizations for relaxation and healing. Her music covers many genres and is a unique blend of musical elements from around the world and spoken meditations, giving the listener a truly unique and relaxing recorded experience. Her recordings are featured on the Monroe Institute's Hemi-Sync label, with binaural sound healing technology added to balance the left and right hemispheres of the brain. Her singing style is a unique immersion into grace with her angelic vocals and sacred songs. Eluv combines her vocals with a tapestry of pure tones of the Crystal Singing Bowls, lush harp strings, and other beautiful sounds. Known for her open-hearted intuitive approach to healing, Eluv's music awakens joy, heart opening, well-being and deep relaxation. She is a featured guest speaker on sound healing and performs live "Sound Healing" concerts globally. Her practice features multiple modalities; Sound Healing with Crystal Singing Bowls, and Angel Harps, Live Healing Music Performances, Composing Music & Guided Meditations, Voice Overs, Mantra, Kirtan, Mandala Art, Essential Oils, Flower Essences. She is a Shamanic practitioner and an Advanced Ancestral Healing practitioner, offering private sessions for people and animals. She can be contacted at eluvmusic@gmail.com